THE SUNKEN
BILLIONS

AGRICULTURE AND RURAL DEVELOPMENT

Seventy-five percent of the world's poor live in rural areas and most are involved in agriculture. In the 21st century, agriculture remains fundamental to economic growth, poverty alleviation, and environmental sustainability. The World Bank's Agriculture and Rural Development publication series presents recent analyses of issues that affect the role of agriculture, including livestock, fisheries, and forestry, as a source of economic development, rural livelihoods, and environmental services. The series is intended for practical application, and we hope that it will serve to inform public discussion, policy formulation, and development planning.

Titles in this series:

Agriculture Investment Sourcebook

Changing the Face of the Waters: The Promise and Challenge of Sustainable Aquaculture

Enhancing Agricultural Innovation: How to Go Beyond the Strengthening of Research Systems

Forests Sourcebook: Practical Guidance for Sustaining Forests in Development Cooperation

Gender in Agriculture Sourcebook

Organization and Performance of Cotton Sectors in Africa: Learning from Reform Experience

Reforming Agricultural Trade for Developing Countries, Volume 1: Key Issues for a Pro-Development Outcome of the Doha Round

Reforming Agricultural Trade for Developing Countries, Volume 2: Quantifying the Impact of Multilateral Trade Reform

Shaping the Future of Water for Agriculture: A Sourcebook for Investment in Agricultural Water Management

The Sunken Billions: The Economic Justification for Fisheries Reform

Sustainable Land Management: Challenges, Opportunities, and Trade-Offs

Sustainable Land Management Sourcebook

Sustaining Forests: A Development Strategy

THE SUNKEN
BILLIONS

The Economic Justification for
Fisheries Reform

THE WORLD BANK
Washington, DC

FAO
Rome

ISBN: 978-0-8213-7790-1
eISBN: 978-0-8213-7914-1
DOI: 10.1596/978-0-8213-7790-1

Library of Congress Cataloging-in-Publication data has been requested.

Cover illustration and design: Critical Stages

CONTENTS

BOXES, FIGURES, AND TABLES

Tables

ACKNOWLEDGMENTS

This study was led by Rolf Willmann (Fisheries and Aquaculture Department, FAO) and Kieran Kelleher (Agriculture and Rural Development Department, World Bank). Ragnar Arnason (University of Iceland) developed the theory and modeling underpinning the study and undertook the economic rents loss calculations. Nicole Franz (FAO) helped with the statistical analyses. The study was undertaken as part of "The Rent Drain" activity of the World Bank's PROFISH Partnership. The study would not have been possible without the support of the Ministry of Foreign Affairs, Iceland, and Agence Française de Développement, France, through their contributions to the initial PROFISH Trust Fund.

The authors are grateful for the contributions provided by John Ward (NMFS), Rashid Sumaila (UBC), and Andrew Smith and Stefania Vannuccini (FAO); and to Ann Shriver (IIFET) and Rebecca Lent (NMFS, moderator) for facilitating special sessions of the Biennial Conference of the International Institute for Fisheries Economics and Trade on "The Rent Drain" in Portsmouth, UK, in 2006 and in Nha Trang, Vietnam, in 2008.

The authors wish to thank the participants in the study design workshops held in Washington, DC and Rome in 2006 for their counsel and advice: Max Agüero, Jan Bojo, Kevin Cleaver, John Dixon, Lidvard Gronnevet, Marea Hatziolos, Eriko Hoshino, Glenn-Marie Lange, Matteo J. Milazzo, Giovanni Ruta, Gert van Santen, Kurt E. Schnier, William E. Schrank, Jon Strand, Laura Tlaiye, John Ward, Ron Zweig; and Serge Garcia, Rognvaldur Hannesson, and John Sutinen.

The authors also wish to express their gratitude for the guidance provided by the concept note peer reviewers: Giovanni Ruta (Environment Department, World Bank), Gert van Santen (consultant), and John Ward (NMFS). The authors are indebted to the insights and encouragement received from commentators and the peer reviewers of the study: Kirk Hamilton (ENV, World Bank), Serge Garcia (consultant), Gordon Munro (UBC), Carl-Christian Schmidt (OECD), Craig Meisner (DECRG, World Bank), Gert van Santen (consultant), and Jon Strand (IMF).

The authors acknowledge the mentoring of Francis (Chris) Christy over the years; the valuable exchanges of views with PROFISH team members in the Sustainable Development Network of the World Bank, including Michael Arbuckle, Lidvard Gronnevet, Marea Hatziolos, Eriko Hoshino, and Oleg Martens; with ARD advisers Chris Delgado, Nwanze Okidegbe, and Cees de Haan; and the logistical support provided by Regina Vasko, Felicitas Doroteo-Gomez, and Joyce Sabaya.

The study was initiated under the guidance of Kevin Cleaver and Sushma Ganguly and completed under the guidance of Juergen Voegele, Director, and Mark Cackler, Manager, of the Agriculture and Rural Development Department of the World Bank.

ABOUT THE AUTHORS

Ragnar Arnason is a leading fisheries economist and a professor in the Economics Faculty at the University of Iceland. He is an expert on the economics of individual transferable quotas (ITQs) used to manage fisheries. ITQs are sometimes considered to be the basis for the economic success of Iceland's fisheries. He has been an advisor to several developing countries on fisheries management issues. He developed the bioeconomic model used to estimate the $50 billion annual loss of potential economic benefits in the global ocean fishery.

Kieran Kelleher is the Fisheries Team Leader in the World Bank's Agriculture and Rural Development Department. He is the manager of the World Bank's Global Partnership on Fisheries—PROFISH—which supported *The Sunken Billions* study. The partnership includes developing countries, leading bilateral donors to the fisheries sector, and technical institutions such as FAO. With a background in biology and economics, he has spent most of his career in developing countries and worked as a fisherman, fish farmer, fisheries scientist, and economic advisor on fisheries to governments. He is the author of global studies on discards, on aquaculture, and on fisheries enforcement.

Rolf Willmann is a Senior Fisheries Planning Officer in FAO's Fisheries and Aquaculture Department. He focuses on policy and management issues in marine capture fisheries, especially on the implementation of the Code of Conduct for Responsible Fisheries. He has led the development of the FAO Ecolabeling Guidelines for Marine Capture Fisheries and undertaken numerous assessments of the economics of capture fisheries in developing countries, with a particular focus on small-scale fisheries and poverty issues. He was the secretary of the Global Conference on Small-Scale Fisheries held in Bangkok in 2008 (http://www.4ssf.org/index.htm).

ACRONYMS AND ABBREVIATIONS

ARD	Agriculture and Rural Development Department, World Bank
DECRG	Development Research Group, World Bank
ENV	Environment Department, World Bank
EU	European Union
FAO	Food and Agriculture Organization of the United Nations
FIES	Fisheries and Aquaculture Information and Statistics Service, FAO
FIEP	Fisheries and Aquaculture Development Planning Service, FAO
GDP	gross domestic product
GRT	gross registered tonnage
GT	gross tonnage
IIFET	International Institute for Fisheries Economics and Trade
IMF	International Monetary Fund
IPCC	Intergovernmental Panel on Climate Change
IPOA	International Plan of Action
ITQ	individual transferable quota
IUU	illegal, unreported, and unregulated fishing
kg	kilogram
kW	kilowatt

MEY	maximum economic yield
MSY	maximum sustainable yield
NMFS	National Marine Fisheries Service (U.S.)
OECD	Organisation for Economic Co-operation and Development
PoI	Plan of Implementation
ton	metric ton (1,000 kg)
UBC	University of British Columbia
WSSD	World Summit on Sustainable Development
WTO	World Trade Organization

All dollar amounts are U.S. dollars unless otherwise indicated.

EXECUTIVE SUMMARY

The contribution of the harvest sector of the world's marine fisheries to the global economy is substantially smaller than it could be. The lost economic benefits are estimated to be on the order of $50 billion annually. Over the past three decades, this cumulative global loss of potential economic benefits is on the order of $2 trillion. The losses represent the difference between the potential and actual net economic benefits from global marine fisheries.

By improving governance of marine fisheries, society could capture a substantial part of this $50 billion annual economic loss. Through comprehensive reform, the fisheries sector could become a basis for economic growth and the creation of alternative livelihoods in many countries. At the same time, a nation's natural capital in the form of fish stocks could be greatly increased and the negative impacts of the fisheries on the marine environment reduced.

In economic terms, some 60 percent of the world's marine fish stocks were "underperforming assets" in 1974, the year when the Food and Agriculture Organization (FAO) initiated its reports on the state of the world's marine fish stocks. By 2004, more than 75 percent of the fish stocks were underperforming, at an estimated annual loss of $50 billion to the global economy. The "sunken billions" is a conservative estimate of this loss. The estimate excludes consideration of losses to recreational fisheries and to marine tourism and losses attributable to illegal fishing are not included. The estimate also excludes consideration of the economic contribution of dependent activities such as fish processing, distribution, and consumption. It excludes the value of biodiversity losses and any compromise to the ocean carbon cycle. These exclusions suggest that the losses to the global economy from unsustainable exploitation of living marine resources substantially exceed $50 billion per year.

For over three decades, the world's marine fish stocks have come under increasing pressure from fishing, from loss of habitats, and from pollution. Rising sea temperatures and the increasing acidity of the oceans are placing further stress on already stressed ecosystems. Illegal fishing and unreported catches undermine fishery science, while subsidies continue to support unsustainable fishing practices.

THE STATE OF MARINE FISH STOCKS AND FISHERIES

The global marine catch has been stagnant for over a decade, while the natural fish capital—the wealth of the oceans—has declined. FAO reports that an increasing proportion of the world's marine fish stocks is either fully exploited or overexploited. Most of the world's most valuable fish stocks are either fully exploited or overexploited. The 25 percent that remains underexploited tends to consist of lower-value species or the least profitable fisheries for such stocks. When fish stocks are fully exploited in the biological sense, the associated fisheries are almost invariably performing below their economic optimum. In some cases, fisheries may be biologically sustainable but still operate at an economic loss. For example, the total catch may be effectively limited by regulations, but in a world of increasing fuel subsidies, the real cost of harvesting the catch may exceed the landed value. The depletion in fish capital resulting from overexploitation is rarely reflected in the reckoning of a nation's overall capital and GDP growth.

This study and previous studies indicate that the current marine catch could be achieved with approximately half of the current global fishing effort. In other words, there is massive overcapacity in the global fleet. The excess fleets competing for the limited fish resources result in stagnant productivity and economic inefficiency. In response to the decline in physical productivity, the global fleet has attempted to maintain profitability by reducing labor costs, lobbying for subsidies, and increasing investment in technology. Partly as a result of the poor economic performance, real income levels of fishers remain depressed as the costs per unit of harvest have increased. Although the recent changes in food and fuel prices have altered the fishery economy, over the past decade real landed fish prices have stagnated, exacerbating the problem. The value of the marine capture seafood production at the point of harvest is some 20 percent of the $400 billion global food fish market. The market strength of processors and retailers and the growth of aquaculture, which now accounts for some 50 percent of food fish production, have contributed to downward pressure on producer prices.

THE ESTIMATE OF THE "SUNKEN BILLIONS"

In technical terms, this study estimates the loss of potential economic rent in the global fishery. For the purposes of this study, economic rent is considered broadly equivalent to *net economic benefits*, which is the term used throughout most of the report. The lost benefits, or the difference between the potential and actual net benefits, can be largely attributed to two factors. First, depleted fish stocks mean that there are simply fewer fish to catch and therefore that the cost of catching is greater than it could be. Second, the massive fleet overcapacity, often described as "too many fishers chasing too few fish," means that the potential benefits are also dissipated through excessive fishing effort.

This study estimated the difference between the potential and actual net economic benefits from global marine fisheries using 2004 as the base year. The estimate was made using a model that aggregated the world's highly diverse fisheries into a single fishery. This made it possible to use the available global fisheries data such as production, value of production, and global fisheries profits as inputs to the model. Some of the global data sets and inputs required for the model are either deficient or less than robust. Consequently, several further assumptions were required, and in each case the rationale behind the assumption is provided. For example, based on available estimates, the maximum sustainable (biological) yield from the world's fisheries was assumed to be 95 million metric tons (tons). To account for the inherent uncertainties in the data and the simplification in the model, estimates of the most likely range of lost economic benefits were obtained using sensitivity analyses and stochastic simulations.

For the base year, 2004, the 95 percent confidence interval for the lost economic benefits in the global marine fishery was found to be between $26 billion and $72 billion, with the most likely estimate to be on the order of $50 billion.

The estimate of $50 billion—the sunken billions—does not take account of several important factors and is thus a conservative estimate of the potential losses. The model does not include the costs of fisheries management and does not reflect the costs to the marine environment and biodiversity resulting from weak fisheries governance. The model does not fully capture the costs of subsidies, or the benefits that would result from efficient fisheries that would favor the least-cost producers. Nor does the model capture the potential downstream economic benefits of more efficient fisheries. The estimate does not count the benefits from recreational fisheries, from marine tourism, or from healthy coral reefs. The estimate is consistent with previous studies, however, and the study provides a replicable and verifiable baseline for future tracking of the economic health of marine fisheries.

The real cumulative global loss of net benefits from inefficient global fisheries during the 1974–2008 period is estimated at $2.2 trillion. To derive the $2.2 trillion value, the estimated loss of $50 billion in 2004 was used as a base value to construct a time series of losses. The 1974–2008 period was used because the FAO produced its first "state of the marine fisheries" report in 1974, the first of a series of 14 such reports. The changing proportion of global fish stocks reported as fully or overexploited in this series was used to build the annual loss estimate. An opportunity cost of capital of 3.5 percent was assumed.

CAPTURING THE SUNKEN BILLIONS

The depletion of a nation's fish stocks constitutes a loss of the nation's stock of natural capital and thus a loss of national wealth. The depletion of global fish stocks constitutes a loss of global natural capital. Economically healthy marine fisheries can deliver a sustainable flow of economic benefits, a natural bounty from good stewardship, rather than being a net drain on society and on global wealth.

Recovery of the sunken billions can take place in two main ways. First, a reduction in fishing effort can rapidly increase productivity, profitability, and net economic benefits from a fishery. Second, rebuilding fish stocks will lead to increased sustainable yields and lower fishing costs. Some fish stocks can rebuild rapidly, but the uncertain dynamics of marine ecosystems means that certain stocks may not be readily rebuilt. One such example is the Canadian cod stocks, which, despite a reduction in fishing effort, have not recovered.

The crisis in the world's marine fisheries is not only a fisheries problem, but one of the political economy of reform. Fisheries reform requires broad-based political will founded on a social consensus. Building such a consensus may take time and may require forging a common vision that endures changes of governments. Experience shows that successful reforms may also require champions or crises to catalyze the process. Fisheries reform will require reduction in fishing effort and fleet capacity. Thus, successful reforms should take the time to build consensus among fishers on the transition pathways, make provisions for creating alternative economic opportunities, establish social safety nets for affected fishers, and generally manage transition in an equitable manner. Successful reforms will require strengthening of marine tenure systems, equitable sharing of benefits from fisheries, and curtailment of illegal fishing. Successful reforms will require reduction or elimination of pernicious subsidies in the transition to sustainability.

Rising food prices and a growing fish food gap for over 1 billion people dependent on fish as their primary source of protein add to the rationale for fishery reform. Rising fuel prices and the need for greater resilience in marine ecosystems in the face of growing pressures from climate change reinforce the arguments for concerted national and international actions to rebuild fish wealth. The heavy carbon footprint of some fisheries and emerging evidence that depletion of marine fisheries may have undermined the ocean carbon cycle add to the justification for fisheries reform. The depletion of global fish stocks cannot, however, be attributed solely to fishing. Pollution, habitat destruction, invasive species, and climate change all play a role in this process.

THE COSTS OF REFORM

Comprehensive reform of marine fisheries governance can capture a substantial proportion of the sunken billions. The transition to economically healthy fisheries will require political will to implement reforms that incur political, social, and economic costs. These are the costs of investing in rebuilding fish stocks, which requires an initial reduction in fishing activity and harvest rates. The benefits of this investment accrue later when fish stocks have grown and when fishing fleets have adjusted. Once recovered, many ocean fisheries can generate a substantial economic surplus and turn a net economic loss to society into a significant driver of economic growth and a basis for alternative livelihood opportunities. However, the social, economic, and institutional costs of this transition must be financed. The allocation of this cost burden between public and private sectors presents challenges both to fiscal policy and management practice.

The most critical reform is the effective removal of the open access condition from marine capture fisheries and the institution of secure marine tenure and property rights systems. Reforms in many instances would also involve the reduction or removal of subsidies that create excess fishing effort and fishing capacity. Reduction or removal of subsidies can, however, cause undesirable economic and social hardship, especially at a time when fishers face volatile prices of fuel and food. Subsidies that create perverse incentives for greater investment and fishing effort in overstressed fisheries tend to reinforce the sector's poverty trap and prevent the creation of economic surplus that can be invested in alternatives, including education and health. The World Bank has suggested that any subsidies should be temporary, as part of a broader strategy to improve fisheries management and enhance productivity. Rather than subsidies, the World Bank has emphasized investment in quality public goods such as science, infrastructure, and human capital, in good governance of natural resources, and in an improved investment climate.

The alternative to reform—business as usual—is a continued decline in global fish wealth, harvest operations that become increasingly inefficient, and growing poverty in fishery-dependent communities. Failure to act implies increased risks of fish stock collapses, increasing political pressure for subsidies, and a sector that, rather than being a net contributor to global wealth, is an increasing drain on society.

THE BIOLOGICAL AND ECONOMIC HEALTH OF FISHERIES

The focus on the declining biological health of the world's fisheries has tended to obscure the even more critical deterioration of the economic health of the fisheries, which stems from poor governance and is both a cause and a result of the biological overexploitation. Economically healthy fisheries are fundamental to achieving accepted goals for the fisheries sector, such as improved livelihoods, food security, increased exports, and the restoration of fish stocks, which is a key objective of the World Summit on Sustainable Development Plan of Implementation. This study makes the economic case for comprehensive reform of fisheries governance and complements ecological and conservation arguments.

Many national and international fishery objectives focus on maintaining or increasing capture fishery production, and it is argued that national policies would benefit from a greater focus on maximizing net benefits and choosing economic or social yield as an objective rather than continuing to manage fisheries with maximum sustainable yield as an objective. Such a socioeconomic focus implies that planners and decision makers devote greater attention to reform of the pernicious incentive structures driving fisheries overexploitation.

A clear picture of the economic health of fisheries is fundamental to building the economic sustainability necessary to conserve and rebuild fish stocks. Such a health check needs to take account of subsidies, environmental externalities, and depletion of fish capital, and underpins any coherent policy debate on fishery reform.

NET BENEFITS AND TENURE

It has long been understood that because the benefits from fish harvests are to individuals but costs of resource reduction are shared, the net benefits from use of common pool resources, such as fish stocks, will tend to be dissipated. In many countries, marine fishery resources are considered to belong to the nation, and governments are charged with stewardship of this public asset. This has in some instances undermined the traditional rights systems observed by local communities and led to a de facto open access condition. The public, or common pool, character of marine fish resources is often deeply embedded in law and practice, so strengthening marine fisheries tenure is a complex undertaking and faces political, social, and legal challenges. It will require good understanding of traditional or de facto fishing rights systems and of the functionality and legitimacy of national fisheries legislation as a basis for bridging the divide between community and national stewardship functions.

It is not the role of this study to be prescriptive with regard to marine fisheries tenure but to raise awareness of the link between tenure and net benefits and to suggest that avoidance of the sensitive issues of marine use rights is likely to result in a continued slide toward poverty for many fishery-dependent communities. Reforms will require empowerment of poor fisher communities, establishment of secure user and property rights, and investment in collective action by a strengthened civil society. In a world of volatile fuel and food prices, any apparent advantage held by small-scale fisheries also needs to be supported by a greater investment in the management of small-scale fisheries.

These are among the many reasons why the economic objectives—increasing the net benefits and wealth from fisheries—need to be at center stage in efforts to resolve the crisis in marine fisheries. Public awareness and understanding of the potential and actual flows of economic benefits can inform the political economy of reform and help leaders move toward socially responsible and sustainable fisheries underpinned by sound scientific advice.

RECOMMENDATIONS

1. Use the results of this study to raise awareness among leaders, stakeholders, and the public of the potential economic and social benefits from improved fisheries governance in contrast to the sector's current drain on society in many countries.
2. Promote country- and fishery-level estimates of the potential economic and social benefits of fisheries reform and assessment of the social and political costs of reform as a basis for national- or fishery-level dialogue.
3. Build a portfolio of experiences in the process of fisheries reform with a focus on the political economy of reform and the design of the reform process, including consideration of the timing and financing of reform and structuring a national dialogue on the reform process. Fisheries reform initiatives should draw on the knowledge and lessons of reforms in other sectors, in particular with regard to the impact on the poor and the effectiveness and equity of adjustment mechanisms.
4. Progressively identify a portfolio of reform pathways based on a consensus vision for the future of a fishery and founded on transparency in the distribution of benefits and social equity in reforms. The common elements of such pathways could include effective stakeholder consultation processes; sound social and economic justifications for change; and an array of social and technical options, including decentralization and comanagement initiatives to create more manageable fishery units. A reform process will bend the trusted tools of fisheries management to new tasks. Sound scientific advice, technical measures such as closed seasons, and effective registration of vessels are likely to form synergies with poverty reduction strategies, transitions out of fisheries, social safety nets, and community comanagement.
5. Review fiscal policies to phase out subsidies that enhance fishing effort and fishing capacity and redirect public support measures toward strengthening fisheries management capacities and institutions, while avoiding social and economic hardships in the fisheries reform process.
6. In an effort to comply with the call of the World Summit on Sustainable Development Plan of Implementation for restoration of fish stocks, countries could, on a timely basis, provide to their public an assessment of the state of national fish stocks and take measures to address the underreporting or misreporting of catches.

Global Trends in Fisheries

1.1 INTRODUCTION

E conomically healthy fisheries are fundamental to achieving accepted goals for the fisheries sector, such as improved livelihoods, exports and food security, and the restoration of fish stocks, which is a key objective of the Plan of Implementation adopted at the World Summit on Sustainable Development in Johannesburg in 2002. Many national and international fishery objectives focus on maintaining or increasing the quantity of capture fishery production while less attention is devoted to the economic health of fisheries.

An analysis of key global trends in fisheries—including fish production and consumption, the state of the fish stocks, and employment in the sector—provides the context and builds a profile of the economic health of the world's marine fisheries. Estimates of the economic value of global marine fishery production and costs of production are used as inputs to an aggregate economic model to derive a range of estimates of potential economic rents lost, largely as a result of suboptimal governance of the marine fisheries worldwide. Key assumptions underlying the model are described.

1.1.1 Purpose and Outcomes of the Study

The purpose of this study is to raise the awareness of decision makers regarding the economic dimensions of the crisis in the world's marine capture fisheries. The target group includes not only fisheries professionals, many of whom grapple with this crisis on a daily basis, but a broader audience of policy and decision makers who can foster reforms in fisheries with a view to rebuilding fish wealth and capital as a basis for economic growth and biologically and economically healthy fisheries.

The study shows that, in aggregate, the global marine fisheries in the base year, 2004, represent a net economic loss to society and often a poverty trap for dependent communities. The study shows that if marine capture fisheries were organized to move fisheries in the direction of maximizing economic efficiency, then national fisheries sectors, fishing communities, and society as a whole would reap substantial economic benefits. The political, social, and economic costs of such reforms are briefly discussed.

1.1.2 Structure of the Study

Chapter 1 provides an overview of trends in global fisheries to set the context for the study.

Chapter 2 describes the economic performance of the world's capture fisheries. The study reviews the main determinants for the economic performance of global fisheries, such as the value of fish production, the cost of factors of production, and productivity trends. The available global data sets are described as a framework for selection of the parameters used in the model. Fisheries are shown to benefit from significant subsidies that often undermine sustainability and maintain inefficiency. Illegal fishing is recognized as a governance failure undermining the economic and biological health of fisheries. Substantial additional work is suggested to remove uncertainties with respect to the magnitude of unrecorded catches at the global level.

Chapter 3 presents the approach and method used to build a bioeconomic model of the aggregate global fishery. Additional technical details of the model are provided in the appendixes.

Chapter 4 presents the results of the analysis, highlighting the poor economic health of the world's marine fisheries and the need for greater attention to improving the economic well-being of fisheries and fishers: as a sustainable source of economic growth, as a pathway out of poverty, as a means of contributing to food security, and as a way to build resilience to the impending effects of climate change.

Chapter 5 discusses the results and draws on available case studies to identify key elements in moving fisheries toward a more economically rational base without sacrificing fundamental social objectives in pursuit of economic efficiency.

Four appendixes provide supplementary information.

1.2 THE DETERIORATING STATE OF THE MARINE FISHERY RESOURCES

The crisis in marine fisheries has been well documented in biological terms. This study focuses on the economic health of the world's fisheries as a complement to the numerous reviews of the ecological state of the global marine fisheries. Globally, the proportion of fully exploited, and either overexploited, depleted, or recovering fish stocks has continued to increase from just above 50 percent of all assessed fish stocks in the mid-1970s to about 75 percent in 2005 (box 1.1) (FAO 2006). This indicates that, in economic terms, more than 75 percent of the world's fisheries are underperforming or are subject to economic overfishing. In 1974, about 40 percent of the assessed stocks were rated as underexploited or moderately exploited. By 2005, this percentage had fallen to 25 percent (FAO 2007c).

Between 1950 and 1970, the recorded catch of both the demersal (bottom-dwelling) and pelagic species (species that live in the upper layers of the sea)

Box 1.1   Stagnating Global Marine Catch

The box figure indicates that the reported global marine catch has stagnated at a level of 80–85 million tons since 1990. This stagnation hides several underlying trends in the composition of the catch as described below.

One-half of the marine capture fish stocks monitored by the FAO are designated as fully exploited, producing at or close to their maximum sustainable yield. Another 25 percent of the marine fish stocks are either overexploited, depleted, or recovering from depletion and are yielding less than their maximum sustainable yield (FAO 2007c). The remaining 25 percent of the marine capture fish stocks are underexploited or moderately exploited, and although this implies that more could be produced, many of these underexploited stocks are low-value species or species for which harvesting may be uneconomical. Global production of seafood from wild stocks is at or close to its long-run biological maximum.

Reported Global Marine Catch, 1950–2006

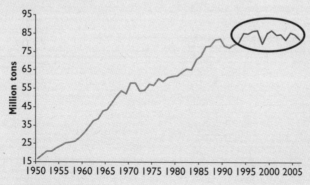

Source: FAO FishStat Plus.

grew considerably (figure 1.1). Since 1970, demersal fish catches have stabilized at around 20 million tons a year, while pelagic catches grew to a peak volume of almost 44 million tons in 1994. Since then, annual pelagic catches have fluctuated between 36 and 41 million tons.

Thus, the global fish supply from marine capture fisheries increasingly relies on lower-value species characterized by large fluctuations in year-to-year productivity, concealing the slow degradation of the demersal high-value resources. About 17 percent of the global catch reported to FAO by member countries is not reported by species group. Thus, the FAO's FishStat Plus database does not readily allow assessment of these species composition changes on a global basis. This change in the species composition of the catch is commonly referred to as "fishing down marine food webs" (Pauly et al. 1998). The stagnant level of production is thus maintained by the relatively higher growth rate of a higher proportion of smaller fish species lower on the food web and a likely decrease in the average age of the catch, which jointly contribute to maintaining fish biomass. In some fisheries, the targets of fishing have also expanded to cover an entire spectrum of species in the ecosystem "fishing through the food webs" (Essington, Beaudreau, and Weidenmann 2006).

The changing patterns of discards (fish caught but dumped unwanted at sea) also suggests that the global catch now comprises substantial quantities of lower-value, previously discarded fish: the amount of fish discarded may have decreased by over 10 million tons between 1994 and 2004 (Kelleher 2005). For example, the quantity of so-called trash fish used for aquaculture feed is estimated to be 5–7 million tons (Tacon 2006; APFIC 2006). There is also growing evidence that the

Figure 1.1   Catch of Selected Species Groups in Marine Fisheries

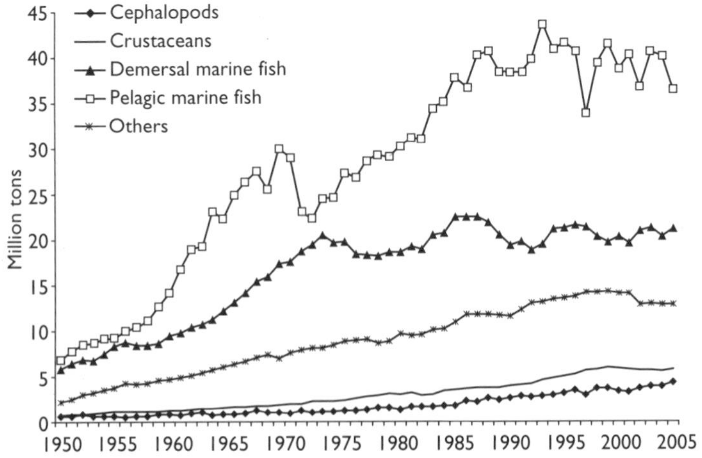

Source: FAO FishStat Plus.

THE SUNKEN BILLIONS

biomass of large predatory fishes has declined substantially from preindustrialized levels in many regions (Myers and Worm 2003; Ahrens and Walters 2005), although this may not hold true for all fisheries (Sibert et al. 2006).

Climatic variability has always been a significant determinant of fish stock growth and decline, and response to variability is part of the daily business of fishing. However, climate change, as described by the Intergovernmental Panel on Climate Change (IPCC 2007), is placing additional stress on fisheries already stressed by pollution, habitat loss, and fishing pressure. Although recent studies on coral reefs (Baird et al. 2007) and reviews of the likely impacts of climate change on the fisheries of the North Atlantic provide important guidance on trends, in the case of developing countries, the impact of changes in sea temperature and ocean acidity on their fish stocks remains largely undetermined. Similarly, the impact of sea-level rise and erratic climatic events on the community and household wealth of coastal fishing populations remains largely unquantified. These added ecological, environmental, and economic stresses caused by climate change add to the urgency and economic justification for restoring the resilience and health of fish stocks (FAO 2008; European Commission 2007; Sustainable Fisheries Livelihoods Project 2007).

1.3 PROFILE AND TRENDS IN GLOBAL FISHERIES PRODUCTION

In 2006, total reported world fishery production (excluding aquatic plants) reached almost 160 million tons (figure 1.2), of which 53 percent originates from marine capture fisheries. Over the past 20 years, the continued growth in world fish production is largely attributable to aquaculture (see figure 1.2).

China is the largest producing country, contributing 49 million tons in 2005, of which 32 million tons are from aquaculture (figure 1.3). Developing countries have contributed more than one-half of total capture fish production since 1990 (figure 1.4) and this share reached more than two-thirds in 2005.

1.4 TRADE AND FISH CONSUMPTION

Rising demand for fish has been a major driver of increased fishing effort. Spurred by the globalization of markets for fish, some 37 percent of global fish production flows into international trade, making fish one of the most traded "agricultural" commodities and accounting for up to 13 percent of global "agricultural" trade. The benefits of increasing globalization in fish trade have nevertheless been reduced by growing overexploitation because ineffective governance of fisheries has allowed the depletion of fish stocks—the natural capital, or fish wealth (ICTSD 2006).

Figure 1.2 World Marine and Inland Capture and Aquaculture Production, 1950–2005

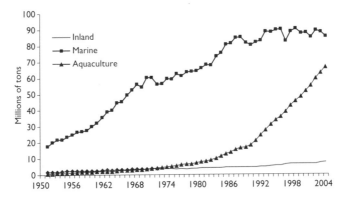

Source: FAO FishStat Plus.

Figure 1.3 World Capture and Aquaculture Production, 1950–2005

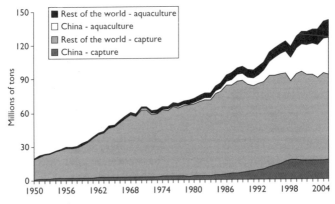

Source: FAO FishStat Plus.

In 2006, total world trade of fish and fishery products reached a record (export) value of $86.4 billion, more than a 10-fold increase since 1976, when global fish trade statistics first became available. The share of developing countries in total fishery exports was 48 percent by value and 57 percent by quantity. Growth in aquaculture production has been an important factor for the global expansion of seafood trade.

The growth in reported global fish production has more than kept pace with population growth (figure 1.5). Based on the reported global fish production, the total amount of fish available for human consumption is estimated to have reached 107 million tons in 2005, providing an average

Figure 1.4 Total Recorded Marine Capture Production by Economic Group, 1970–2005

Source: FAO FishStat Plus.

Figure 1.5 World Population and Global Fish Supply, 1970–2003

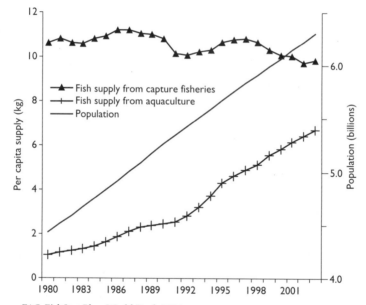

Source: FAO FishStat Plus; World Bank 2006.

global per capita fish supply of 16.5 kilograms, but with large differences across regions and countries as well as within countries (FAO 2007c). These global values, however, may not adequately reflect important subsistence fish consumption and consumption of unreported commercial production from small-scale fisheries.

Figure 1.6   Regional Trends in Annual Fish Supply, 1961–2003

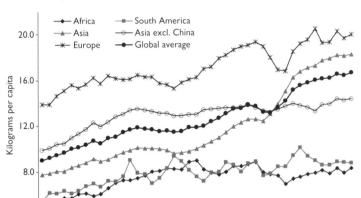

Source: FAO FishStat Plus; FAO 2007a.

Aquaculture products continue to capture an increasing share of global markets for fish. This is driven by technological advances in production, relatively lower production costs (compared with capture fisheries), and globalization of fish trade. The competition from aquaculture places additional economic stress on capture fisheries and contributes to trade disputes as farmed fish capture market share from traditional producers.

Rising demand in China and Europe has largely driven the increase in average global per capita fish consumption (figure 1.6). This global increase was particularly pronounced in the 1980s and 1990s but has since stabilized at around 16 kilograms per capita per year (FAO 2007a). Per capita consumption of fish in South America is stabilizing after a peak in 1995, but per capita consumption in Africa and South America remains low (see figure 1.6). In both regions, but especially in sub-Saharan Africa, low animal protein intake is believed to be largely a result of low per capita incomes. Traditionally, low-value fish and fishery products provide cheap protein to the poorer populations in these regions and in Asia. Africa is the only continent where per capita fish consumption has been in decline (less than half the global average), and because fish tends to be the lowest priced animal protein, this trend raises concern about the nutritional quality of the diet, particularly in sub-Saharan Africa. Aquaculture production has responded to the increasing demand in Asia. However, despite recent growth, African aquaculture has been unable to respond to the continent's nutritional needs. The increased demand for aquaculture and livestock feeds based on trash fish and low-value species has a potential negative impact on the availability and accessibility of these products for direct human consumption (APFIC 2006).

CHAPTER TWO

The Economic Performance of World Marine Capture Fisheries

The economic performance of global marine capture fisheries is determined by the quantity of fish caught, the price of fish, the harvesting costs, and the productivity of the fisheries. The following sections summarize the global profile of each of these determinant factors and discuss the closely related issues of subsidies and excess capacity in the global fishing fleet.

2.1 VALUE OF PRODUCTION AND GLOBAL FISH PRICES

In 2004 (the base year for the study), the total nominal value[1] of reported global fish production was estimated at $148 billion, of which capture fisheries was $85 billion and aquaculture was $63 billion. The total estimated value of the reported marine catch of 85.7 million tons was $78.8 billion (FAO 2007c).[2]

2.1.1 Ex-vessel Prices

The nominal average ex-vessel price was $918 per metric ton for the reported marine catch and $666 per ton for the reported inland (freshwater) catch. The average farmgate price for cultured fish was $1,393 per ton. The higher unit price for aquaculture products is a result of the production of high-value species (for example, shrimp and salmon). The ex-vessel prices are considered to be conservative and close to true market prices, being relatively free of taxes, subsidies, and other market-distorting influences.

2.1.2 Export Prices

Fish price data sets are relatively incomplete at the global level: the primary long-term price data series is the fish export unit value derived from FAO's FishStat Plus trade statistics. The unit value of exports may underestimate the global trend in real fish prices. On the one hand, higher-value fish products tend to be exported. On the other hand, aquaculture has a growing share in world fish trade, and prices of many cultured species have tended to decline from the initial elevated price levels.

Because of the changing product composition of exports, the export values are only indicative of the price trends, but the values nevertheless show several interesting features (see figure 2.1). There was a significant decline in fish prices between 1978 and 1985, followed by a strong price rise from the mid-1980s to the early 1990s, a gradual decline until 2001, and a recovery in prices during the most recent years. The real unit value of exports in 2004 was no higher than it was in the late 1980s. This strongly suggests that the global price of fish in 2004 was not significantly different from that in the late 1980s.

Setting aside numerous supply-driven real price fluctuations, the real prices of many fish commodities saw little change between the early 1990s and late 2007 (Josupeit 2008, Asche and Bjørndal 1999). The notable exceptions are increased fish meal and fish oil prices, which have been driven by higher demand for meat and aquaculture products. Tuna prices and some whitefish

Figure 2.1 Trends in the Nominal and Real Unit Export Value of Fishery Products

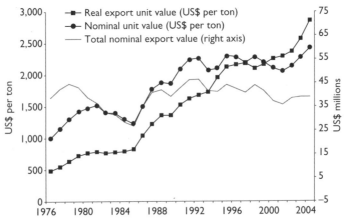

Source: FAO FishStat Plus.
Note: The deflator used for real values is the U.S. producer price index for all commodities, base year 1982 (Delgado et al. 2003). Values exclude aquatic plants.

prices have also increased, while supplies from aquaculture have dampened prices for some products. Fillet and product yields have improved, wastage has been reduced, and supply chains shortened, making downstream industry increasingly more efficient and competitive, often decreasing profit margins for primary producers and intermediaries.

Thus, although the unit value of the aggregate reported catch has remained relatively constant, the higher proportion of relatively lower-value "trash fish" and small pelagic species is buoyed up by the increasing scarcity value of species higher on the food web, such as lobster or grouper. The scarcity of some higher-value species has created opportunities to fish in deeper waters, often at a higher cost per unit of catch and also at a cost to the relatively unknown bio-diversity of the continental slopes.

Growth in demand for fish is concentrated in developing countries where populations and per capita incomes show strong growth. However, survey data from China in the period 1980–2000 indicate only slight increases in the real price of fish (Delgado et al. 2003). Recent studies show substantial increases in Chinese seafood consumption with increases of over 100 percent in lower-income households to over 150 percent for higher-income families between 1998 and 2005 (Pan Chenjun 2007). In contrast, while demand continues to grow in the United States and real prices of fresh fish show a long-term increasing trend, the price of the traditional frozen products and particularly of canned products has declined during the last 30 years (figure 2.2). More

Figure 2.2   Trends in U.S. Real Price Indexes for Fish and Seafood Products, 1947–2006

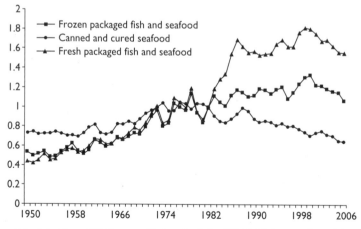

Source: Calculated from U.S. Bureau of Labor Statistics 2007. This is an update of figure 3.2 in Delgado et al. (2003).
Note: 1982 is the base year ("1" on the y-axis).

recently, a weakening U.S. dollar exchange rate and consumer spending may be contributing to a recent decline in U.S. shrimp imports, a key seafood indicator (Seafood International 2008).

2.1.3 Value of Intangibles

Healthy marine ecosystems generate a range of intangible values, which are difficult to estimate in the absence of robust global data sets and agreed valuation methods. These values arise from marine biodiversity, the existence value of megafauna, and the value of environment services from natural assets such as healthy reefs (Cesar 2000; UNEP-WCMC 2006; Worm et al. 2006). There may be additional potential benefits from ocean carbon sequestration resulting from healthy fish stocks (Lutz 2008). The global fishing fleet currently has substantial excess capacity, but a global fleet that is "in balance" with the fish stocks can significantly reduce the carbon footprint of the industry. The bioeconomic model used in this study does not include a valuation of these intangibles.

2.2 FISHING COSTS AND PRODUCTIVITY

There is no representative global data set on the costs of fishing. However, costs and earnings studies are available from a number of countries and fisheries. Fishing costs vary greatly by type of fishery and locality: for example, trawl fisheries tend to have high fuel costs, while many smaller vessels are nonmotorized and the cost of subsistence fishing may be little more than the cost of the labor involved. In general, the major cost factors for most fisheries are:

- labor (30–50 percent of total costs);
- fuel (10–25 percent);
- fishing gear (5–15 percent);
- repair and maintenance (5–10 percent); and
- capital cost, such as depreciation and interest (5–25 percent).

The trends in the costs of each of these factors of production are relevant not only for an understanding of the historical trends in fisheries but also to provide a basis for future projections of, for example, the effect of changing fuel prices. Available cost data must be treated with some caution, because the true cost data tend to be confounded by taxes and subsidies. There is ample evidence that at the global level productivity has deteriorated, especially in recent years, as the majority of producers incur higher fishing costs while the global catch has remained stagnant.

2.2.1 Fuel Prices and Productivity

The cost of crude oil not only directly relates to fishing fuel costs but also indirectly affects the cost of fishing nets and lines and of vessel construction and repair. Figure 2.3 shows an index of the real price of crude oil and an index of the real material costs in U.S. ship building. For comparison purposes, the index of the real unit value of fish exports is also illustrated. It shows that until about mid-1980, real unit export value rose faster than crude oil and unit material costs, but since the late 1980s price and cost trends have been fairly similar, with the crude oil price climbing steeply since 2000.

Since 2000, fuel subsidies have probably played an increasingly important role in supporting the financial viability of fishing operations in some countries. Such fuel subsidies (mostly forgone taxes) to the fishing sector by governments globally are estimated to be in the range of $4.2 billion to $8.5 billion per year (Sumaila et al. 2008).

In the absence of productivity gains, figure 2.3 strongly suggests that the economic performance of global marine fisheries is unlikely to have improved since the early 1990s. Several factors continue to undermine productivity. These include rising oil prices; rising costs of fishing gear and vessels, often compounded by unfavorable exchange rates (for countries that import factors

Figure 2.3   Real Trends in Crude Oil Price, Vessel Material Costs, and Fish Export Unit Value (1998 = 100)

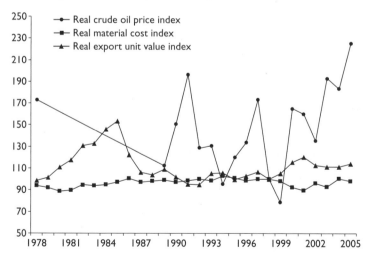

Source: FAO FishStat Plus; FAO FIEP; U.S. Department of Energy, Energy Information Administration; http://www.coltoncompany.com/shipbldg/statistics/index.htm, based on data from the U.S. Bureau of Labor Statistics. Deflator used for real values: U.S. producer price index for all commodities, base year 1982 (Delgado et al. 2003).

of production); an increasing regulatory burden; and depletion of inshore stocks, causing fishers to travel farther to fishing grounds.

By contrast, nonmotorized fisheries, fisheries that use passive gears (such as traps) and thus relatively less fuel, and fisheries that have ready access to export markets may have seen an improvement in profitability in this period. Technology also has driven productivity gains. Using sophisticated fish-finding equipment, tuna purse seiners in the western Indian Ocean can now harvest three times the annual catch of seiners operating in the mid-1980s. New designs of trawls reduce the required engine power and fuel consumption by a factor of 33 percent or more (Richard and Tait 1997). Electronic sale of fish while vessels are still at sea reduces transaction costs, helps prevent loss of product quality and value, and makes markets more efficient (Jensen 2007). However, as these innovations are adopted and spread throughout a fleet, aggregate productivity falls, and the economic rents generated through the increasing productivity are not maintained.

Fuel consumption varies considerably, depending not only on different fishing methods and types of fisheries but also on the fuel efficiency of engines. At the global level, on average each ton of fish landed required nearly half a ton of fuel. In value terms, production of a ton of fish worth $918 required $282 worth of fuel, or 31 percent of the output value, in 2004. The impact of the recent (2007–08) doubling of fuel prices is briefly addressed in a subsequent section and the overall trend in fish, food, and fuel prices is illustrated in figure 2.4.

Figure 2.4   Trends in Fish, Food, and Fuel Prices

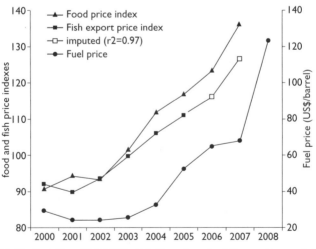

Source: FAO FishStat Plus; FAO FIEP; U.S. Department of Energy, Energy Information Administration. The imputed fish price index for 2006 and 2007 was derived from a correlation with the FAO Food Price Index.

2.2.2 Trends in Employment, Labor Productivity, and Fishing Incomes

During the past three decades, the number of fishers and fish farmers has grown at a higher rate than the world's population growth rate (figure 2.5). Catching and fish-farming activities provided livelihoods to an estimated 41 million people in 2004 employed as either part-time or full-time fishworkers.[3] Applying an assumed ratio of 1:3 for direct employment (production) and secondary activities (postharvest processing, marketing, distribution), respectively (FAO 2007b), about 123 million people are estimated to be involved in postharvest processing, distribution, and marketing activities. Many countries do not separate capture fisheries and aquaculture employment data. Based on available fisheries labor statistics, the number of capture fishers account for three-quarters of employment in fisheries globally.

Although employment in capture fisheries has been growing steadily in most low- and middle-income countries, fisheries employment in most industrial economies has been declining. This decline can be attributed to several factors, including the relatively low remuneration in relation to often high-risk and difficult working conditions, growing investment in labor-saving onboard equipment (FAO 2007c), and a failure to attract younger workers. The increase in numbers of fishworkers in developing countries is not only a result of increased fish production activities. For some communities, fisheries is a growing poverty trap and, in the absence of alternatives, a livelihood of last resort.

Asia has by far the highest share and growth rate in the numbers of fishers and fish farmers (figure 2.6). In this region, the number of fishers increased threefold over the three decades from 1970 to 2000—reflecting both a strong increase in

Figure 2.5   Global Population Growth and Trend of Total Number of Capture Fishers

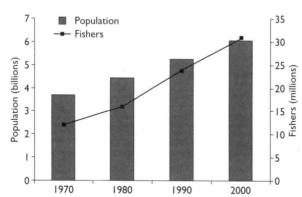

Source: FAO 1999 (1970, 1990 data); FAO 2007c (1990, 2000 data); FAO FIES.

Figure 2.6   Total Number of Capture Fishers by Region

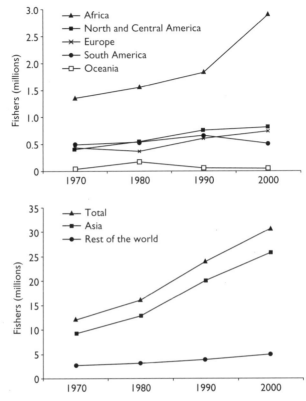

Source: FAO 1999 (1970, 1990 data); FAO 2007c (1990, 2000, 2004 data).

part-time and occasional employment in capture fisheries and the growth in aquaculture activities. In Africa, growth was more moderate until 1990 but has accelerated sharply since then.

An indicator of labor productivity is the output per person measured either in physical or value terms. Figure 2.7 shows the average output per fisher valued at average ex-vessel prices in 1998–2000. Average output per fisher ranged from a high of just above $19,000 in Europe to about $2,231 in Africa and $1,720 in Asia, about a 10-fold difference.

The low labor productivity in Africa and Asia reflects low fishing incomes in most countries in these regions. For example, the estimated average gross revenue per full-time fisher in India's marine fisheries was $3,400 in 2004. Small-scale fishers grossed an average of $1,870, and the figure for fishers on

Figure 2.7   Gross Revenue per Marine and Inland Capture Fisher
(average 1998–2000 in US$)

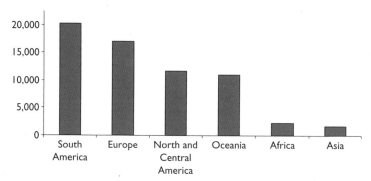

Source: FAO 2002, 2007c; FAO FishStat Plus.
Note: Data for South America have been adjusted to take low-value fish used for reduction (fishmeal) into account.

industrial vessels was $5,490 (Kurien 2007). Average labor productivity is higher when only full-time fishers are considered, but labor productivity is still significantly below labor productivity values in other primary sectors of these economies.

There is both hard and anecdotal evidence of low levels of crew remunerations in many of the world's marine fisheries. For example, Vietnamese workers on fishing vessels operating in African waters and flagged in countries with per capita income some 10 times higher than in Vietnam receive a monthly pay of $150–180 and working conditions include 16- to 18-hour work days.[4] A significant share of crews on Thai industrial fishing vessels are from Myanmar and Cambodia, two countries with widespread poverty and average incomes some eight times lower than those of Thailand. Based on average country poverty data, some 5.8 million, or 20 percent, of the world's 29 million fishers, may be small-scale fishers that earn less than $1 a day (FAO 2004).

The strong growth in capture fisheries employment (that is, fishers operating full time, part time, occasionally, or with unspecified status) has not resulted in a commensurate increase in inland and marine capture fisheries production. As shown in figure 2.8, the average reported harvest per capture fisher has declined by 42 percent, from more than 5 tons annually in 1970 to only 3.1 tons in 2000.

The significance of this decline in average output per fisher has to be seen in the context of the enormous technological developments that have taken place in the world's capture fisheries during this period, including large-scale motorization of traditional small-scale fisheries, the expansion of active fishing techniques such as trawling and purse seining, the introduction of

Figure 2.8   Annual Catch (Marine and Inland) per Capture Fisher, 1970–2000

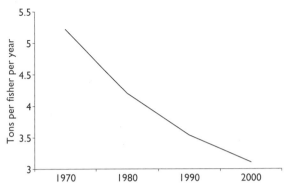

Source: FAO FishStat Plus; FAO 2002; FAO 2007c; FAO FIES.

increasingly sophisticated fish-finding and navigation equipment, and the growing use of modern means of communication. Although this technological progress has certainly increased labor productivity in many fisheries, at the aggregate global level the resource constraint in combination with widespread open access conditions have prevented an increase in average labor productivity in the world's capture fisheries. Overall, productivity has significantly declined, a decline caused by a shrinking resource base and a growing number of fishers.

Because the number of fishing vessels has also increased significantly over the last several decades (see below), at the global level the productivity-enhancing investments in capture fisheries have on average yielded poor returns and have stymied growth in labor productivity and incomes in the sector.

2.3   FISHING EFFORT AND FISHING FLEETS

Fishing effort is a composite indicator of fishing activity. It includes the number, type, and power of fishing vessels and the type and amount of fishing gear. It captures the contribution of navigation and fish-finding equipment, as well as the skill of the skipper and fishing crew. Effective effort is difficult to quantify even in a single fishery, and there is considerable uncertainty about the current level of global fishing effort. Given the multiple dimensions of fishing effort, it is understandable why no global statistics are available.

The primary factor influencing fishing effort is the size of the global fishing fleet, characterized in terms of vessel numbers, tonnage, and engine power, and type of fishing gear as described in the following section.

In biological terms, fishing effort equates with fishing mortality. The functional relationship is determined by a factor known as the "catchability coefficient." This coefficient is a measure of both the level of harvesting technology and fishing skill as well as the relative ease of harvesting the fish stock in terms of its distribution and abundance. This variable is captured in the bioeconomic model by the schooling parameter discussed in chapter three.

2.3.1 Development in the Global Fishing Fleet

The reported global fleet has increased numerically by about 75 percent over the past 30 years to a total of approximately 4 million decked and undecked units in 2004 (FAO 2007c; figure 2.9). The number of decked (motorized) vessels more than doubled in this period, and the average age of the global fleet of

Figure 2.9   Total Number of Decked and Undecked Fishing Vessels Per Region, 1970–98

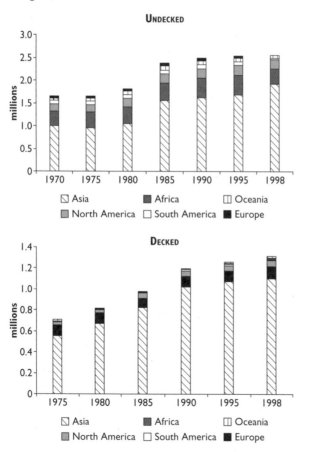

Source: FAO FIES; FAO 1999.

large fishing vessels has continued to increase. Asia accounts by far for the highest number of vessels, both decked and undecked.

FAO data on national fishing fleets are primarily derived from administrative records, which may not always be current. For example, national records may include fishing vessels that are not currently operational and they frequently omit large numbers of unregistered small-scale fishing vessels (FAO 2007c). A further difficulty in maintaining a consistent data set results from the progressive change in the measurement of vessel size from gross registered tonnage to gross tonnage and the reflagging of vessels to flags of convenience.

For large vessels, the Lloyd's database (http://www.lrfairplay.com/) of vessels provides a relatively robust global data set for fishing vessels above 100 gross tons. However, coverage is incomplete. Although FAO fleet statistics show an increase in global fleet size since the early 1990s, the Lloyd's register shows a decline in recent years in the number of fishing vessels larger than 100 gross tons (figure 2.10). This divergence in trends can be explained partly by the evolution of the Chinese fleet, which is incompletely listed in the Lloyd's Register because it is domestically insured. The FAO statistics used for this fleet and for smaller vessels have been compiled from national statistics. In 2002, China adopted a five-year program to reduce its commercial fleet by 30,000 vessels (7 percent) by 2007. However, the numbers of commercial fishing vessels reported to FAO in both 2003 and 2004 are above the number reported as being in operation in 2002 (FAO 2007c).

Figure 2.10   Estimated Number of New Fishing Vessels Built and Total Registered Fleet Size (Vessels over 100 GT/GRT)

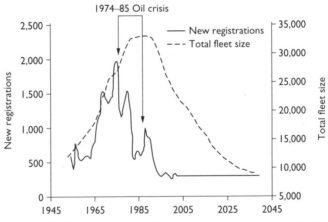

Source: Based on Lloyd's data for vessels 100 gross tons or more and reproduced with permission; S. Garcia and The Royal Society (Garcia and Grainger 2005, fig. 11, p. 29).

2.3.2 Development in Fishing Capacity and Fleet Productivity

Fishing capacity is the amount of fishing effort that can be produced in a given time by a fishing vessel or fleet under full utilization for a given fishery resource condition (FAO 2000). Both the increase in vessel numbers and in vessel technology have enhanced the capacity of the global fleet and facilitated access to an expanding range of marine fishery resources and more efficient use of these resources.

Fitzpatrick (1996) estimated that the technological coefficient, a parameter of vessel capacity, grew at a rate of 4.3 percent per year.[5] Assuming that this trend has continued, growth in technological efficiency coupled with growth in the number of vessels suggests a steeply rising global fleet capacity. The capacity index shown in figure 2.11 is a multiple of the total number of decked vessels and the technological coefficient.[6] The trend line of the catch/capacity index demonstrates that the global harvesting productivity has on average declined by a factor of six.

The exploitation of a growing number of less productive fish stocks partly explains this decline in harvesting productivity, but the buildup of fishing overcapacity is clearly a major contributing factor. Thus, the gains from technological progress have generally not been realized because the limited fish stocks require a concomitant reduction in the number of vessels to allow for improved vessel productivity.

The decline in physical productivity is compounded by the decreasing spread between average harvesting costs and average ex-vessel fish prices,

Figure 2.11 Evolution of Global Fleet Productivity (Decked Vessels)

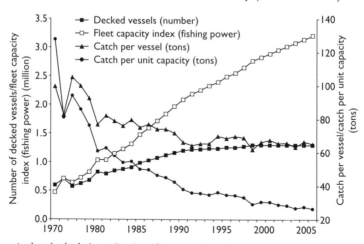

Source: Authors' calculations; Garcia and Newton (1997); FAO FishStat Plus; FAO FIEP.

causing depressed profit margins and reinvestment. Although this has a dampening effect on growth in fleet capacity, depressed fleet reinvestment may retard a shift to more energy-efficient harvesting technologies and a reduction in the carbon footprint of the fishing industry.

Many countries have adopted policies to limit the growth of national fishing capacity, both to protect the aquatic resources and to make fishing more economically viable for the harvesting enterprises (FAO 2007c). These policies have proven difficult and costly to implement in many instances, and even when numbers of vessels have been successfully reduced (Curtis and Squires 2007), the reduction in fishing effort has been considerably less than proportional, because it is the less efficient vessels that tend to exit the fishery, and expansion in technical efficiency counters the reduction in vessel numbers.

The global fleet has attempted to maintain its profitability in several ways: by reducing real labor costs, by fleet modernization, and by introduction of fuel-efficient technologies and practices, particularly in developed countries. Vessels are also reported to remain in harbor for increasingly longer periods of the year, focusing harvesting on peak fishing seasons.

The receipt of government financial support has also assisted both vessel operators and crews, for instance, through income compensation for crews. Subsidies in the world's marine fisheries have received growing attention in recent years and are further discussed later.

2.3.3 The Effects of Changing Fuel and Food Prices

The impact of higher fuel and food prices on marine capture fisheries is becoming clearer. The effect depends on the interplay between the impact of the fuel price change on the level of fishing effort; the price elasticity of demand for fish in economies in which the cost of the entire food basket increases; and the changes in per capita incomes that underlie the demand for fish. The outcome of this interplay is likely to be specific to the economy of individual fisheries and the markets for the products of that fishery (table 2.1).

Table 2.1 The Effects of Fuel and Food Price Increases	
Fuel price increases may:	**Food price increases may:**
• reduce fishing effort as a result of higher costs • reduce fish supply and drive fish prices higher • change fishing patterns to less fuel-intensive modes • result in higher fuel subsidies	• increase fish prices to more than compensate for higher harvest costs • redirect forage fisheries (fish meal) catches to higher-value human food products • allow aquaculture products to permanently capture market share from marine capture fishery products • stimulate increased fishing effort

Source: Authors.

A number of fuel-intensive fleets ceased to operate in mid-2008; others are benefiting from subsidized fuel to stay operational. The past trend to replace labor with capital is likely to slow or reverse as labor-intensive fisheries become relatively more viable. Products from less fuel-intensive aquaculture may also capture markets. Reduced fishing effort is likely to result in recovery of some fish stocks. Meanwhile, the economic hardship may offer an opportunity for measures to bring fishing capacity into balance with resources.

2.4 SUBSIDIES

Many subsidies in the fisheries sector are pernicious because they foster overcapacity and overexploitation of fish stocks. By reducing the cost of harvesting, for example, through fuel subsidies or grants for new fishing vessels, subsidies enable fishing to continue at previously uneconomic levels. Subsidies effectively counter the economic incentive to cease fishing when it is unprofitable (box 2.1).

Box 2.1 What Are Subsidies?

There is a wide range of definitions of subsidies. The most precise is probably that used by the World Trade Organization (Article 1 of the Agreement on Subsidies and Countervailing Measures), which can be summarized as follows: a financial contribution by the public sector that provides private benefits to the fisheries sector. The contribution can be direct or indirect (such as forgone tax revenue). The contributions can be provided as goods or services, or as income or price supports. Subsidies exclude provision of general infrastructure, or "purchases goods."

Common fisheries sector subsidies include grants, concessional credit and insurance, tax exemptions, fuel price support (or fuel tax exemption), direct payments to industry, such as vessel buyback schemes, fish price support, and public financing of fisheries access agreements. In addition, subsidies have variously been considered to include government fisheries extension and scientific research services. Policy changes, such as relaxation of environmental regulations governing fisheries or special work permits for migrant fishworkers (crew), can also reduce costs in the sector and such distortions also have been regarded as a form of subsidy.

The justification offered for subsidies ranges from protection of infant industries, through national food security and prevention of fish spoilage, to social rationales such as preservation of traditional livelihoods and poverty reduction.

Fuel subsidies are an example of a transfer that reduces the cost of fishing. The reduced costs restore profitability and create perverse incentives for continued fishing in the face of declining catches. The result is overfishing, fleet overcapitalization, reduced economic efficiency of the sector, and resource rent dissipation.

Source: Authors; Schrank 2003; WTO 1994.

Table 2.2 Estimate of Fisheries Subsidies with Direct Impact on Fishing Capacity per Year, 2000 ($ billion)

Subsidy types	Developing countries	Developed countries	Global total	Percent of global total
Fuel	1.3	5.08	6.4	63.5
Surplus fish purchases	0	0.03	0	0.3
Vessel construction, renewal and modernization	0.6	1.30	1.9	18.9
Tax exemption programs	0.4	0.34	0.7	7.3
Fishing access agreements	0	1.00	1.0	9.9
Global total	2.3	7.75	10.05	100

Source: Compiled from Milazzo 1998, with updated information from Sumaila and Pauly 2006; Sharp and Sumaila forthcoming; and Sumaila et al. 2007.

Several direct estimates of subsidies and financial transfers to the fisheries sector have been made (Millazo 1998; Pricewaterhouse Coopers 2000; OECD 2000; Sumaila and Pauly 2006), and several attempts have been made to classify fisheries subsidies in relation to their perceived impact on the sustainability of fisheries and on international trade (for example, "traffic lights," as proposed by the United States to the WTO (World Trade Organization) Negotiating Group on Rules. Recent discussions also have focused attention on both the social rationale and potential negative impacts of subsidies to small-scale fishing (Schorr and Caddy 2007). An updated global estimate of capacity-enhancing subsidies for both developing and developed countries is shown in table 2.2.

Over $10 billion in subsidies that directly impact fishing capacity and foster rent dissipation were provided in 2000. Close to 80 percent of the total global subsidy is provided by developed countries. Transfers of public funds and supports to the fisheries sector are directed at a spectrum of goods ranging from the purely public to the purely private. The issue of subsidies is closely linked to the policies and principles underlying fiscal regimes for fisheries, which must untangle the web of weak property rights prevalent in most fisheries. The issue of subsidies is further addressed later in the discussion. Subsidies are not distinguished as a separate input to the bioeconomic model used to estimate the sunken billions.

2.5 THE COSTS OF FISHERY MANAGEMENT

Both fishers and the public sector incur fisheries management costs. The costs to the public are significant, ranging from 1 to 14 percent of the value of landings for enforcement (monitoring, control, and surveillance) activities alone

(Kelleher 2002a) and imposing a substantial burden on international fisheries management processes (High Seas Task Force 2006). The generation of scientific advice and the process of management also represent significant costs (Arnason, Hannesson, and Schrank 2000).

The public costs of fisheries management have not been taken into account in the estimate of lost rents. These costs are not included in the global bioeconomic model because representative global data are deficient and because the relationship between expenditures on fisheries management and net benefit from the fishery remains unclear. The few studies that have been made of fisheries management costs in developing countries suggest inadequately low levels of management expenditures (Willmann, Boonchuwong, and Piumsombun 2003).

2.5.1 Costs Associated with Illegal, Unreported, and Unregistered Fishing

The International Plan of Action to combat illegal, unreported, and unregistered (IUU) fishing (FAO 2001a) bundles these three related activities and, as a result, studies have tended to bundle rather than disaggregate estimates of the economic impact of these quite distinct fishing activities. Illegal and unreported fishing are of particular interest for the estimate of rents. However, to account for the economic impacts of illegal and unreported fishing, greater knowledge on the scale of both and a greater understanding of the economics of illegal fishing is required (Sutinen and Kuperan 1994; OECD 2006; Sumaila, Alder, and Keith 2006; MRAG and UBC 2008).

The estimates of unreported fishing, or more specifically of underreported or misreported catches, are of considerable interest for the purposes of assessment of economic benefits from fishing. By definition, such estimates are not reflected in FAO's FishStat Plus. The estimates range from multiples of national FishStat Plus values, for example, in the case of some countries that underreport catches from highly dispersed small-scale fisheries to deliberate underreporting of 10–20 percent or more in managed fisheries where fishers seek to circumvent quota restrictions. However, in the absence of a robust basis for adjusting the reported catch to the estimated real catch, the FAO FishStat Plus values remain as the core global data set used in the global bioeconomic model.

Illegal fishing can be considered as additional effort that takes place at a lower cost than legitimate effort. However, the production from this illegal effort may be recorded or included in the estimates of catches, or landings. For example, the catch from use of an illegal type of net may be indistinguishable from that of a legal net. Illicit catches affect rent generation by undermining the governance structure of the fishery, by undermining market prices for legitimate product, and by imposing added management enforcement costs as indicated earlier.

Illicit catches are frequently unreported—for example, fish under a legal size limit, or catch in excess of quota. The resulting inaccuracies in catch statistics are an important source of uncertainty with respect to scientific advice on fisheries management (Pauly et al. 2002; FAO 2002; Kelleher 2002b; Pitcher et al. 2002; Corveler 2002), and the depletion of many stocks has been attributed partly to the inaccuracy of the historical catch data. The parallel markets for illicit fish set a discounted price for fish, not only directly through illicit landings but also by avoidance of sanitary controls or rules of origin regulations, such that normally compliant fishers may be compelled to revert to illicit practices to remain solvent.

NOTES

1. Nominal value is the value of money in different years; real value adjusts for differences in the price level in those years.
2. Estimate provided by FAO Fisheries and Aquaculture Information and Statistics Service (FIES). All values exclude marine plants. The unit values from "FAO World Fishery Production Estimated Value by Species Groups" were weighted by the quantity of the respective marine catches in 2004. Discards are assumed to have zero value.
3. Preliminary results of a new World Bank–FAO–WorldFish Center study indicate that this may be a substantial underestimate.
4. The International Labour Organization of the United Nations recently adopted a comprehensive new labor standard, the "Work in Fishing Convention," which will come into effect when ratified by 10 of the ILO's 180 member states, of which at least 8 are coastal states.
5. For 13 different vessel types (from pirogues of 10 meters up to super trawlers of 120 meters), the coefficient increased on average from 0.54 in 1965 to 1.98 in 1995, or by about 366 percent in 30 years.
6. In some managed fisheries, increase in technological capacity has been limited by gear regulations and other fishery management measures.

CHAPTER THREE

Estimate of Net Economic Loss in the Global Marine Fishery

3.1 BACKGROUND

This study draws on previous efforts to develop an economic assessment for the world's marine capture fisheries. Christy and Scott (1965) suggested that the growth of marine fisheries production would stagnate and proposed that the "maximize sustained yield" objective be replaced by a "maximize rent from the sea" objective. In 1992 (revised, 1993) FAO estimated the aggregate operating deficit incurred by the world's fishing fleets at $54 billion in 1989, the base year of the study (see box 3.1). A study by Garcia and Newton (1997) indicated that an economically efficient global capture fishery required a reduction of between 25 percent and 53 percent in the global fishing fleet.

Because of the deficit of information on the economic health of the world's fisheries, the 2005 World Bank report "Where Is the Wealth of Nations?" was unable to take specific account of fisheries. To address this deficit in the knowledge of the global fishery economy, a workshop was held under the auspices of the World Bank's PROFISH Program (Kelleher and Willmann 2006). The workshop also recognized the need to highlight the current level of global economic rents loss and to raise awareness on economic objectives of fisheries management.

The workshop identified two alternative approaches to the task. One approach is to estimate the rents and rents loss in each of the world's fisheries or a representative sample of them. This is a major undertaking.

Box 3.1   The Framework of Prior Studies

The 1992 FAO study "Marine Fisheries and the Law of the Sea—a Decade of Change" (revised, 1993) estimated the aggregate operating deficit incurred by the world's fishing fleets at $22 billion for the base year of the study (1989). If the cost of capital cost is added, aggregated deficit was estimated at $54 billion per year, or nearly three-fourths of the estimated gross revenue of $70 billion from the global marine fish harvest. The primary causes of these deficits were attributed to the open access management regime that governed most of the world fisheries and rampant subsidization of the global fishing fleet.

Building on the FAO study, Garcia and Newton (1997) examined the trends and future perspective of world fisheries. The authors confirmed the broad conclusions of the 1992 FAO study, the large overcapacity of the global fishing fleet, and the need to reform fishery management systems if long-term economic and environmental sustainability of the world fishery system was to be achieved. They concluded that even though the world's oceans seemed to be exploited at maximum sustainable yield levels, an economically efficient global capture fishery would require either a 43 percent reduction in global fishing costs, or a 71 percent global price increase of capture fishery products, or a global capture fleet capacity reduction between 25 percent and 53 percent.

An alternative, simpler approach is to regard the global ocean fishery as one aggregate fishery. This approach has several advantages. The data requirements are immensely reduced. Many of these global fisheries data are readily available, and the model manipulation and calculations are a fraction of that required for a study of a high number of individual fisheries. The aggregate approach, regarding the global fisheries as a single fishery is considered the only way to quickly and inexpensively obtain reasonable estimates of the global fisheries rents loss in a transparent and replicable manner.

On this basis, the workshop recommended that two independent approaches to the estimation of the loss of economic rents in global marine fisheries be prepared. Each estimate would serve as a cross-check on the other:

- The first study would estimate the global rent drain (or potential loss of net benefits) through an aggregate model of the global fishery. This report documents the results of this first approach.
- The second, companion study would undertake a set of case studies on economic rents in a representative set of fisheries and endeavor to extrapolate results and lessons of case studies to the global level. This work is still in progress.

The study reported here is based on a simple aggregative model for the global fishery. It improves on the previous FAO studies mentioned above in at least three important ways. First, the concept of fisheries rents and rents loss is made explicit. Second, the theoretical assumptions, the assumptions based on empirical data, and the way the conclusions are derived are clearly and systematically specified. This allows testing, improvement, and updating. Third, the study systematically accounts for a level of uncertainty in the estimated values and assumptions. This is done in two ways: first, a standard sensitivity analysis of the calculated rents loss to the basic input data for the global aggregated fishery model provides upper and lower bounds on the rent loss estimates. Second, reasonable probability distributions for the basic input data for the model are assumed and the resulting probability distribution of the calculated rents loss derived. On this basis, statistical confidence intervals are produced for the rent loss estimate using stochastic (Monte Carlo) simulations.

3.2 USE OF THE TERMS *NET BENEFITS* AND *ECONOMIC RENTS*

Economists traditionally use economic rents as a measure of the net economic benefits attributable to a natural resource. Rents are not equal to profits—the difference is fixed costs and so-called intramarginal profits. However, rents and profits are usually similar and may sometimes be identical. The economic performance of the global marine fisheries may be measured as the difference between maximum rents obtainable from the fisheries and the actual rents currently obtained.

This estimate of the loss of fisheries rents in global marine capture fisheries focuses on the harvesting sector, that is, the fishery up to the point of first sale. An economically efficient fishery up to the point of first sale will also drive additional downstream efficiencies, for example, in fish processing. This is because, to be efficient, the harvesting sector will adjust the quantity, quality, and timing of landings to the demand from downstream sectors. Estimates of rents from such potential downstream efficiency gains are not captured in the model presented here but are briefly addressed in the subsequent discussion.

In this study, the terms *net benefits, economic rents,* and *rents* are equivalent, and these terms are used interchangeably in the text. In the pure economic sense, however, they are not equivalent. Box 3.2 and appendix 1 describe these concepts in more technical detail.

As already mentioned, this study estimates this loss of potential economic benefits, or rent dissipation, at an aggregate global level. The global level of rent dissipation is an excellent (inverse) metric of the economic and biological health of the global fishery. The economic objective is to maximize the net economic benefits (sustainable rents) flowing from the fishery. For the great majority of

Box 3.2   Net Benefits, Economic Rents, and Overfishing

The resource rent is a measure of the net economic benefits from the harvest of wild fish stocks. Different fisheries generate different levels of resource rent. For example, a fishery for a high-value species in coastal waters (which has a low cost of harvesting) will generate more rent (or profits to fishers) than a fishery for a low-value species harvested at high cost in deep water. As more fishers join a profitable fishery, they add to the aggregate costs of catching the limited quantity of fish available. As a result, the aggregate net benefit, or economic rent, decreases, becoming dissipated among the fishers in the form of higher costs and lower returns for their fishing operations or fishing effort. The rents may even become negative when public financial transfers or subsidies are provided to support an economically unhealthy fishery. As more fishers make greater efforts (for example, by fishing longer hours or investing in more fishing gear) to maintain their previous profits or catch levels, the fishers tend to deplete the fish stock capital that sustains the productivity of the fishery. This further reduces the potential net benefits.

As soon as the level of fishing effort moves above the point of maximum economic yield, a situation of economic overfishing exists. Such economic overfishing can exist even if the fish stock itself remains healthy or biologically sustainable. This is illustrated in the box figure.

Maximum Sustainable Yield (MSY) and Maximum Economic Yield (MEY)

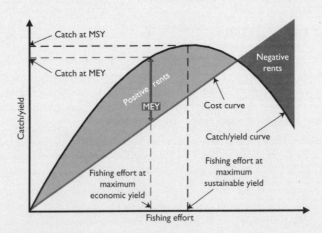

Economists traditionally measure the net economic benefits from a natural resource such as a fish stock by economic rents. Rents are not equal to profits but are usually similar, and may sometimes be identical, to profits. Thus, the inefficiency of fisheries may be measured as the difference between maximum rents obtainable from the fisheries and the actual rents currently obtained.

commercial fish stocks, this implies a biomass level in excess of the one producing the maximum sustainable yield. Even in fisheries with quite high discount rates and comparatively low biomass growth rates, the biomass level at which the economic rents are maximized almost always exceeds the biomass level that can provide the maximum sustainable physical yield (Grafton, Kompass, and Hilborn 2007). Thus, as a general rule, economically healthy fisheries require biologically healthy fish stocks, while biologically healthy fish stocks do not necessarily mean economically healthy fisheries.

3.3   DESCRIPTION OF THE AGGREGATE MODEL

Based on Arnason 2007, an aggregate model of the global fisheries is specified to estimate rents loss for the global marine fishery. This model and procedure for fitting the model are detailed in appendix 2. The model entails several gross abstractions from the real world. In particular, the model assumes that global fisheries can be modeled as a single fish stock with an aggregate biomass growth function. Similarly, the global fishing industry is represented by an aggregate fisheries profit function, composed of an aggregate harvesting function, relating the harvest to fishing effort and biomass, and an aggregate cost function relating fishing effort to fisheries costs.

Fisheries and the rents they generate are dynamic and rarely in equilibrium, implying several approaches for calculating rents losses. This study compares maximum sustainable rents to the actual rents in the base year (2004). The difference is taken to represent the rents loss in the base year. In this study, sustainable (or long-run) rents are identical to profits, so that maximum sustainable rents (MSR) are obtained at the fishing effort level corresponding to the maximum economic yield (MEY) (see figure in box 3.1). The rents loss estimate assumes that the existing biological overfishing is entirely reversible in the long run. Finally, the estimate does not take account of the costs of restoring the global fishery to economic health.

Treating the diverse global fisheries as a single aggregate fishery allows for a model with a manageable number of parameters. A set of available observations on the global fisheries is used to estimate the parameters.

The model's simplifications and uncertainty with respect to global fisheries parameters are partially offset by sensitivity analysis of the results and stochastic simulations to establish reasonable upper and lower bounds and confidence limits for the global fisheries rents loss. It is anticipated that the model will be further tuned and cross-checked using a series of case studies currently in preparation.

3.3.1   Schaefer and Fox Models

The population dynamics of the exploitable aggregate biomass (the global fishery) are modeled through a logistic, or Schaefer-type, model and through a Fox

Figure 3.1   Comparative Yield-effort Curves Corresponding to the Logistic (Schaefer) and Fox Biomass Growth Functions

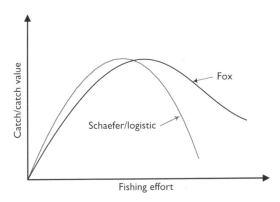

Source: Authors' depiction.

model. The main difference between these two biomass growth functions is that the Fox model assumes that, all else being equal, the biomass is much more resilient to increasing fishing effort, in other words, the sustainable biomass and harvest will decline more slowly as fishing effort increases (figure 3.1). The Fox model is consistent with the experience from the global fishery that even though many of the most valuable demersal fish stocks have become depleted, the aggregate global harvest has continued to increase and has not contracted significantly despite ever-increasing fishing effort.

These models were selected because they are in widespread use for fisheries assessment. Two models are used to reflect the uncertainty with regard to the shape of the biomass growth function of the global aggregate fishery, which may not necessarily equate to a simple sum of individual fishery functions. Other types of aggregate biomass growth functions exist of course. Many of them, however, fall within the range defined by the logistic and the Fox functions.

The shape of the yield-effort curve is given principally by the carrying capacity, or pristine state, of the fish stock(s), the maximum sustainable yield, and the parameters of the harvesting (catch production) function. Of these parameters, estimates of the maximum sustainable yield are more robust than estimates of the other two parameters because comprehensive global marine fish catch statistics are available for over 50 years and harvest trends have been relatively stable for nearly two decades in the range of 79–88 million tons.

3.4   MODEL PARAMETERS AND DATA

As noted earlier, this study assumes that global fisheries can be modeled as a single fish stock. Recovery of lost rent also assumes that biological overfishing is reversible. The basic data used to estimate model parameters and parameter assumptions are listed in table 3.1. The sources for the data and justification for assumptions are provided in the following sections. Further details and the theoretical relationships are further explained in appendix 2. The year 2004 is taken as the base year for the model because several robust data sets are available for that period. However, adjusted data from other years, or a series of years, is used where data for 2004 are deficient.

3.4.1 Global Maximum Sustainable Yield and Carrying Capacity

The global MSY is assumed to be higher than the reported marine catch in the base year (85.7 million tons, FAO FishStat Plus) plus estimated discards (7.3 million tons), which gives a total of 93 million tons. A conservative value of 95 million tons is used in the model. Though higher than the catch in the base year, this value is lower than the sum of the maximum reported catch for each species group in the past (101 million tons) (FAO FishStat Plus). It is also in the same range as that suggested by Gulland in 1971 (100 million tons) and lower than a maximum of 115 million tons suggested by Christy and Scott (1965).

This estimate of the global MSY refers to conventional fisheries only. For example, Antarctic krill is the subject of increasing attention as new harvesting technologies develop and markets for Omega 3 fish oils expand. A major expansion of this fishery could substantially raise the global MSY.

Since the 1990s, reported marine catches have fluctuated between 79 million and 86 million tons without an apparent trend. Given the estimate of the MSY,

Table 3.1 Model Empirical Inputs and Estimated Parameters		
Model inputs	**Model input values**	**Units of measurement**
Biological		
Maximum sustainable yield	95	Millions of tons
Global biomass carrying capacity	453	Millions of tons
Biomass growth in 2004	−2	Millions of tons
Fishing industry		
Landings in 2004	85.7	Millions of tons
Value of landings in 2004	78.8	$ billions
Fisheries profits in 2004	−5	$ billions
Other parameters		
Schooling parameter	0.70	No units
Fixed cost ratio in 2004	0	No units
Elasticity of demand with respect to biomass	0.2	No units

Source: See following sections.

this suggests that the current global fishery is now located to the right of the MSY (see figure in box 2.2). This means that current global fish stocks are smaller than those corresponding to the MSY, in accordance with the general belief that the global fishery is biologically overfished.

The carrying capacity corresponding to the equilibrium MEY is assessed as 453 million tons. This is based on the average relationship between the known carrying capacity and the MSY for a number of fisheries (see table A4.2 in appendix 4).

3.4.2   Biomass Growth in the Base Year

The stability of the aggregate reported catch over the recent past is consistent with a relatively constant aggregate global biomass. During this period, some stocks (for example, demersal stocks such as cod and hake in parts of the Atlantic) have declined markedly in response to fishing pressure, climatic factors, and other influences. Other stocks, such as some pelagic stocks in the North Atlantic, have increased while other large stocks have remained largely unchanged (FAO 2005). Overall, it appears unlikely that in the base year, 2004, there was a significant net increase or decline in global stocks of commercial marine species. However, because global reported catches in 2004 were close to the upper boundary of annual global catches since the 1990s and reported catches in 2005 were lower, it is conservatively assumed, that in 2004 global marine commercial biomass growth was negative, or −2 million tons.

3.4.3   Volume of Landings in the Base Year and Reported and Real Marine Fisheries Catches

In accordance with official FAO statistics (FAO FishStat Plus), the global catch in the base year is taken to be 85.7 tons. Acknowledging the deficiencies of the FAO FishStat Plus records, the FAO has repeatedly called for more comprehensive and accurate reporting of fish catches (FAO 2001a). The level of acknowledged mis- and underreporting of catch has been addressed with varying degrees of success by different authors. The reasons for misreporting vary widely from deliberate underreporting of quota species and deficiencies in transmission of information to FAO, to widespread underestimates of small-scale fisheries production, and to reported overestimates of fish production in the case of China and possibly in other countries. The estimated level of underreporting varies widely. The estimates of underreporting range from 1.2 to 1.8 times the catch (as reported to FAO) in reportedly well-managed fisheries to several times the reported catch in countries with extensive and isolated small-scale fisheries or with high levels of illegal fishing (Oceanic Développement 2001; Kelleher 2002b; MRAG and UBC 2008; Zeller and Pauly 2007; Pauly 1995; Watson and Pauly 2001). However, in the absence of a robust basis for adjusting the reported catch to the estimated real catch, the FAO FishStat Plus values remain as the core data set for this study.

3.4.4 Value of Landings in the Base Year

The value of landings in 2004 is discussed extensively in chapter 2. Based on published production value data and other information, it is estimated that this value was $78.8 billion (FAO 2007c), which corresponds to an average landed price of $0.918 per kilogram.

3.4.5 Harvesting Costs

As indicated in chapter 2, the estimate of harvesting costs must be treated with due caution because of the weak and incomplete data on the world's fishing fleets. The data sets used include:

- A robust set of fleet and productivity data for 21 major fishing nations that contribute about 40 percent to global marine capture production (see table A4.1 in appendix 4).[1] These data are biased toward industrial fisheries but are considered to be representative of all industrial fisheries.
- Sample cost and earnings data for the European fleets (EU 25). The data refer to the fleets of 20 countries that contributed 6.8 percent to the global fish harvest in 2004 (Salz 2006; Concerted Action 2004).
- A recent set of cost and earnings data for India's industrial and small-scale fisheries (Kurien 2007). These fisheries contribute about 2.5 percent to global marine fish harvest. This data set has been taken to represent tropical developing countries' fisheries.

Cost of fuel

Fuel consumption and costs are estimated on the basis of the vessel and engine horsepower data of the fleets listed in table A4.1 in appendix 4. It is assumed that the average vessel activity is 2,000 hours per year and the average world market price of diesel fuel is $548 per ton.[2] Fuel consumption and costs are raised to the global level on a pro rata basis of the contribution of these fleets to global catches. The result is an estimated annual global fuel consumption of 41 million tons valued at $22.5 billion.[3] This decrease in the fuel consumption of the global fishing fleet, compared with the previous estimate (46.7 million tons valued at $14 billion in 1989 prices [FAO 1993]), results from a relatively constant number of fishing vessels above 100 gross tons in the Lloyd's database and from a reduction in overall tonnage from about 15 million gross tons in 1992 to 12.6 million gross tons in 2004. Fuel efficiency has also improved in some fleets and closed seasons may have reduced fishing time.

Cost of labor

The 1993 FAO study based its labor cost estimate on a total number of employed crew of 12.98 million and an average annual crew income of $1,749, leading to an estimated total labor cost of $22.7 billion. The growth in the numbers of fishers, including part-time and occasional fishers, since 1992 suggests that total labor cost of the global fishing fleet has increased. However, labor productivity in terms of catch per fisher and catch value per fisher has decreased. Working hours have increased and safety at sea has deteriorated (ILO 2000), making fishing the profession with the highest labor mortality rate. However, the deterioration in working conditions is not necessarily reflected in labor costs. It is concluded that real per capita crew remuneration has declined and that global labor cost has remained at a relatively constant nominal level of $22.7 billion per year.

Costs of other factors of production
Total operating costs (average 2002–04) exclusive of fuel and labor costs of the fleets from 20 EU countries amounted to $292,000 per 1,000 kilowatt engine power (Salz 2006).[4] Applying this value to the fleets of the 21 fishing nations listed in table A4.1 gives annual operating costs of $13.97 billion (exclusive of fuel and labor). These fleets contribute about 40 percent to world harvest, which would make the estimated global total $34.9 billion. However, these operating costs are lower in small-scale fisheries in developing countries. In India, for example, the operating costs (excluding fuel and labor) in small-scale marine fisheries are on average $90 per ton of fish landed (Kurien 2007). Assuming that small-scale fisheries contribute about 25 percent[5] to the global marine catch and that the cost structure of the remaining 75 percent of fisheries is accurately represented by the 21 fleets referenced above, the global estimate for these other operating costs is $28.1 billion.

This estimate is consistent with the comprehensive costs and earnings data compiled for the European fleet (Salz 2006). However, it is substantially lower than the cost of comparable items indicated in the FAO (1993) study (a total of $55.9 billion—maintenance and repair, $30.2 billion; supplies and gear, $18.5 billion; and insurance, $7.2 billion). The FAO estimates are higher largely because they are based on percentages of the vessel replacement costs and on vessels normally insured and subject to regular surveys. Many fishing vessels do not fall in this category, especially small-scale fishing vessels both in developed and in developing countries.

Cost of capital
The estimate is based on the comprehensive costs and earnings data set available for the European fishing fleet. A capital value per unit of vessel power (kilowatt) was applied to the fleets of 21 fishing nations in the European Union (see table A4.1 in appendix 4). This value was then raised to the global total by dividing by the ratio of the contribution of these fleets to the world marine fish harvest, resulting in a value of $127 billion for total fleet investment.[6]

Total capital costs were conservatively calculated at 8.3 percent of the capital value of the fleet. This resulted in total capital costs of $10.5 billion. Depreciation of this capital was conservatively calculated at 4.3 percent per year,

Table 3.2 Estimated Capital Cost of Global Fishing Fleet *(US$ billion)*

Category	1993 FAO study	Current estimate
Total fleet investment	319.0	127.0
Depreciation cost	—	5.4
Interest cost	—	5.1
Total cost of capital	31.9	10.5

Source: Authors' calculations.
Note: — Data are not available.

resulting in global fishing fleet depreciation of $5.4 billion. Interest costs were calculated at 4 percent, which is an estimate based on secure long-term U.S. dollar investments such as 30-year U.S. treasury bonds. Total estimated capital costs are summarized in table 3.2. For comparison purposes, total capital costs according to the FAO 1993 study are also listed.

The estimate for the total capital invested in the fleet given in the 1993 FAO is higher than the current estimate because it was based on the estimated replacement value (FAO 1993). However, this value ($319 million) is considered an overestimate because the method was applied in the absence of knowledge of the age structure of the fleet and the market prices of vessels at the time.

3.4.6 Profitability

The world's fishing fleet is estimated to have had an operating profit of $5.5 billion in 2004. However, the fleet incurred a capital cost estimated at $10.5 billion, which puts the global fisheries profitability into negative territory, with an estimated deficit of $5 billion in 2004, the base year (table 3.3). These estimates are net of financial subsidies; that is, subsidies have already been subtracted.

It should be noted that profit estimates for the global fishing fleet suffer from a scarcity of reliable fleet cost and earnings data. Many countries do not systematically collect fisheries cost and earnings or profitability data, and these data are particularly deficient for small-scale, artisanal, and subsistence fishing. Even when such data are collected, fishers are often reluctant to provide complete and accurate information, and available information is often distorted by subsidies or taxes. Although based on limited samples, there are nevertheless indications that substantial numbers of fisheries are unprofitable or experience declining profitability (Lery, Prado, and Tietze 1999; Tietze et al. 2001; Tietze et al. 2005; Watson and Seidel 2003; Hoshino and Matsuda 2007).

Fishing that operates at a real economic loss is unlikely to continue without subsidies or forms of vertical integration that capture downstream value. This further narrows the possible range of values for global fleet and fishing profits.

Table 3.3	Global Fleet Estimated Profits, Current and Previous Studies	
Category	1993 FAO study	Current estimate
Value of catch	70	78.8
Fuel costs	14	22.5
Labor costs	22.7	22.7
Other operating costs	55.9	28.1
Operating profit or loss	−22.6	5.5
Total cost of capital	31.9	10.5
Global fleet profitability (deficit)	−54.4	−5.0

Source: Author's calculations; FAO 1993 (base year 1989).

In addition, the "tragedy of the commons" suggests that where forms of open access persist (which is the case in many of the world's fisheries), profits will be dissipated. The value of landings and costs of many factors of production are often known. This again narrows the range for the estimate of profits.

3.4.7 Schooling Parameter

Harvests of species with a strong tendency to congregate in relatively dense schools or shoals (such as herrings, anchovies, and sardines) are often little influenced by the overall biomass of the stock (Hannesson 1993). The opposite is true for species that are relatively uniformly distributed over the fishing grounds (such as cod or sharks). For these species, harvests tend to vary proportionately with the available biomass for any given level of fishing effort.

The schooling parameter reflects these features of fisheries and normally has a value between zero and unity. The lower the schooling parameter, the more pronounced the schooling behavior and the less dependent the harvest is on biomass. For many commercial species (for instance, many bottom-dwelling, or demersal, species and shellfish), the parameter would be close to unity (Arnason 1984). For pelagic species (such as tuna, herring, or sardine), it is often much lower (Bjorndal 1987). A schooling parameter of less than unity leads to a discontinuity in sustainable yield and revenue functions. These discontinuities are of concern because they correspond to a fisheries collapse if fishing effort is maintained above that level for some time.

In the harvesting function for the global fishery, the aggregate schooling parameter should reflect the schooling behavior of the different fisheries. An average of schooling parameters by fishery groups weighted by their maximum sustainable yield levels gives an aggregate schooling parameter of approximately 0.7, which is the value used in this study (see table A4.3 in appendix 4).

3.4.8 Elasticity of Demand with Respect to Biomass

In the global fisheries model employed in this study, the average price of landings depends on the global marine commercial biomass according to a coefficient referred to as the elasticity of demand with respect to biomass. The model uses a value of 0.2 for this parameter, which means that if the global biomass doubles, then the average price of landing increases by 20 percent. The coefficient and the value of the coefficient are based on following rationale.

Fishing activities initially target the most valuable fish stocks and the most profitable fisheries. These high-value species tend to be (but are not always) those high in the marine food chain. As the fishing effort increases, the most valuable stocks become depleted and the fishing activity targets less valuable fish stocks (or in some cases operates in deeper waters on the continental slopes) or targets species at lower trophic levels. This is known as "fishing down and through the food webs." In this situation of overfishing, the higher proportion of lower-value species tends to depress the average price of the aggregate catch.

However when the reverse takes place, under a governance regime that restores biomasses and the health of fish stocks, the average price will tend to rise. However, this generalization must be qualified in terms of the trophic level of the target species and efforts to achieve ecosystem balance across related fisheries. If the target species is a high-value prey species (such as shrimp), then rebuilding the stock of predators (in this example, lower-value fish at a higher trophic level that eat shrimp) may in fact reduce the average price of the aggregate landings (Hannesson 2002). Nevertheless, in general, as stocks rebuild there will tend to be more, larger fish in the catch. Larger fish are generally (but not always) more valuable, which results in a higher average price for the global catch.

Under an effective fisheries management system, the unit price of landed fish usually increases, sometimes substantially (Homans and Wilen 1997; Homans and Wilen 2005). For example, in individual-transferable-quota-based fisheries (one of many choices for improved fisheries management), the average price of landings increases substantially compared with the price before introduction of the ITQ scheme (Herrmann 1996). The reasons include more selective fishing practices, better handling of caught fish, and better coordination between demand for fish and the supply of landings. The increased price is not necessarily related to the more valuable composition of the catch referred to earlier. Finally, there is growing evidence that heavily fished resources are less stable (Anderson et al. 2008), so stock recovery is likely to stabilize supplies and prices and improve the efficiency of harvesting.

3.4.8 The Fixed Cost Ratio

In this study, the loss of potential rents is estimated as the difference between rents in the base year and maximum sustainable rents, that is, maximum rents where biomass (the fish stock) and the capital stock (fleet) are in equilibrium. This equilibrium prevails when fish stocks have been rebuilt and when the fleet has fully adjusted to sustainable catch levels. During the period of fleet adjustment, or long-run economic change, the capital costs, normally regarded as fixed costs, are actually variable. Therefore, for the purposes of comparing base year and maximum sustainable rents, all costs are considered variable costs, and for these theoretical reasons, the fixed cost ratio is set to zero in these calculations. This does not mean that capital costs are ignored in this study but that, for the purposes of the rents loss calculation in this study, they are regarded as variable.

3.4.9 Management Costs and Subsidies

As explained earlier, the costs of fisheries management are not included in the bioeconomic model. Nor are subsidies separately identified in the cost estimates. The existence of subsidies reduces the observed costs so the reported deficit may be underestimated. These additional factors underline the conservative nature of the rents loss estimate.

NOTES

1. The countries are China, EU-15, Iceland, Japan, Republic of Korea, Norway, and the Russian Federation.
2. The impact of recent changes in fuel prices is discussed elsewhere in this report.
3. Taking as a basis data from more than 250 fisheries and spatially resolved catch statistics for 2000, Tyedmers, Watson, and Pauly (2005) estimated global fuel consumption at almost 50 billion liters, equal to 42.5 million tons. On the basis of country-by-country fishing fleet data, Smith (in press) estimated global fuel consumption at 38 million tons.
4. Averages of 2002–04 have been used and converted into US$ at an exchange rate of 1 euro = US$1.107.
5. Because the production from small-scale fisheries tends to be underestimated, or underreported, this value may be an underestimate. Chuenpagdee et al. 2006 suggest that 25 percent may be a minimum value. Current work in progress by the FAO and the WorldFish Center under the World Bank's PROFISH Program (the "Big Numbers" project) also confirm that production from small-scale fisheries may be substantially underestimated.
6. The use of EU cost data may overestimate capital cost because of the presumed higher capital intensity of EU fishing fleets. However, a comparison with Kurien's marine capture data set for India comprising primarily small-scale and semi-industrial fishing fleets suggest that this is not the case. Capital investment per unit of harvest show comparatively similar values: world (based on EU data) $1,494 per ton; and India $1,240 per ton. In the case of depreciation costs, these were estimated even higher, on average, in Indian than in EU marine fisheries.

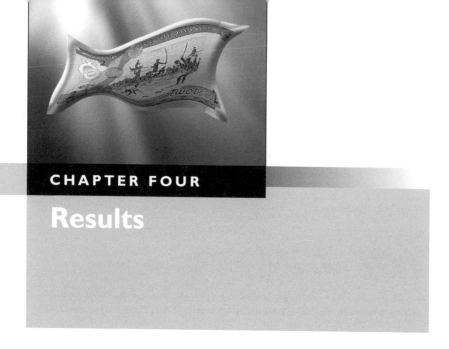

CHAPTER FOUR

Results

4.1 MAIN RESULTS

The loss of net benefits, expressed as forgone rents, is estimated to be on the order of $50 billion in 2004, the base year. Because of model and input data limitations, this estimate is best considered as the most probable value of a range of possible values. Specifically, the most probable point estimate of the global fisheries rent loss is $51 billion with an 80 percent confidence level that the value is between $37 billion and $67 billion.

The rents loss estimate ranges between $45 and $59 billion in the base year, depending on whether the underlying biomass growth function applied is the Schaefer logistic or the Fox function. Table 4.1 summarizes the main results of these calculations for the two biomass growth functions. The Fox biomass growth function estimates a higher current fisheries rents loss primarily because the current level of overexploitation is substantially greater when the Fox function applies. A priori, there is no reason to choose one biomass growth function above the other and the point estimate of $51 billion assumes an equal probability of each function applying.

Based on the loss of net benefits in 2004, the real cumulative global loss of wealth over the past three decades is estimated at $2.2 trillion. This estimate is made by assuming a linear relationship between the rents and the state of the world's fish stocks as reported by FAO at various intervals since 1974. The estimated rents loss in the base year is projected from 1974 to 2007 and raised on the basis of the changing percentage of global fish stocks, reported by FAO as

Table 4.1   Main Results—Point Estimates of Rents

Category	Units	Current Logistic	Current Fox	Optimal Logistic	Optimal Fox	Difference Logistic	Difference Fox
Biomass	Million of tons	148.4	92.3	314.2	262.9	165.8	170.6
Harvest	Million of tons	85.7	85.7	80.8	81.6	−4.9	−4.1
Effort	Index	1.00	1.00	0.56	0.46	−0.44	−0.54
Profits	US$ billions	−5.000	−5.000	39.502	54.035	44.502	59.035
Rents	US$ billions	−5.000	−5.000	39.502	54.035	44.502	59.035

Source: Authors.

fully or overexploited. A conservative opportunity cost of capital of 3.5 percent is assumed. Details of the estimate are provided in table A4.5 in appendix 4.

To maximize sustainable rents from the global fishery, the model indicates that fishing effort should be reduced by 44 to 54 percent depending on whether the aggregate global commercial fishery biomass growth is better described by the logistic or the Fox biomass growth function. The models indicate that biomass levels more than double in the case of the logistic and triple in the case of the Fox biomass growth function compared with the base year estimates. In both cases, sustainable marine fishery harvests are reduced by about 4 million tons compared with the base year harvest.

A summary of the results of the sensitivity analysis and the confidence intervals for the rents loss estimate is provided in section 4.5 below.

4.2   EVIDENCE FROM GLOBAL STUDIES

Although this study is not directly comparable with previous studies, all studies (table 4.2) carry the same message: at the aggregate level, the current annual net benefits from marine capture fisheries are tens of billions of U.S. dollars less than the potential benefits. Society continues to be a net contributor to the global fisheries economy through depletion of the national and global fish capital and through subsidies.

4.3   EVIDENCE FROM CASE STUDIES

A range of case studies strongly indicates the potential for substantial increases in rents and net benefits from fisheries. The different approaches to estimating current and potential rents or similar indexes of net benefits preclude a synthesis of all the available studies in a coherent manner as part of this study.[1] However, table 4.3 and table A4.4 in appendix 4 demonstrate that potential rents range from a significant fraction of the current fishery revenues to multiples of

Table 4.2 Estimates of the Economic Losses from Global Marine Fisheries

Source	Estimate of losses	Drivers/focus of proposed solutions
FAO 1993	$54 aggregate loss, or approximately 75 percent of the gross revenue	Open access, subsidies
Garcia and Newton 1997	$46 billion deficit	Overcapacity, loss of high-value species
Sanchirico and Wilen 2002	$90 billion (future projection)	Rents in ITQ fisheries approach 60–70 percent of gross revenues.
Wilen 2005	$80 billion	Secure tenure
World Bank (this study)	$51 billion	Comprehensive governance reform

Table 4.3 Illustrative Rents Losses in Major Fisheries Assessed with the Model Used in This Study

Fishery	Base year	Base year harvest (1,000 tons)	Base year revenues (US$ millions)	Rents loss as percentage of revenues
Vietnam Gulf of Tonkin demersal multigear	2006	235	178	29
Iceland cod multigear	2005	215	775	55
Namibia hake demersal trawl	2002	156	69	136
Peru anchoveta purse seine	2006	5,800	562	29
Bangladesh hilsa artisanal multigear	2005	99	199	58

Source: Selected case studies in progress FAO/World Bank (see appendix A4.4).

the current fishery revenues. Several fisheries managed in a scientific and responsible manner may yet continue to underperform with regard to rent generation (Kirkley et al. 2006). For example, the potential economic benefits from rebuilding 17 overfished stocks in the United States is estimated at $567 million, or approximately three times the estimated net present value of the fisheries without rebuilding (Sumaila and Suatoni 2006). In a followup to this study, rents loss estimates for a representative range of fisheries will help tune the global rents loss estimate and raise stakeholder awareness on the potential net benefits from improved governance in specific fisheries.

4.4 LINKS TO THE BROADER ECONOMY

The fisheries rents that are generated may be invested in productive physical, human, or social capital, and the net gains from these investments can subsequently be reinvested. Thus, generation of fisheries rents allows fishing economies to choose a higher economic growth path. For countries that are highly dependent on fisheries, harnessing the potential economic growth effects of fisheries rationalization can substantially improve general economic welfare.

4.4.1 Contributions to Economic Growth and GDP

The upstream and downstream economic links, or "multiplier effects," add significantly to the contribution of the fishing industry to the GDP and to wealth creation, because the fishing industry is a base industry that supports economic activity in other sectors of the economy including services (Arnason 1995; Agnarsson and Arnason 2007). In addition, the fishing industry is a disproportionately strong foreign exchange earner in many developing countries, and to the extent that the availability of foreign currency constrains economic output, the economic benefits from the sector may be greater than is apparent from the national accounts. For example, the contribution of the fishing industry in the Pacific Islands has been estimated to be some 30 percent higher than is usually presented in national accounts (Gillett and Lightfoot 2001; Zeller, Booth, and Pauly 2006). An efficient and stable harvest subsector is the basis for maintaining the sector's contribution to GDP.

The study has focused on the marine fisheries to the point of landing, or first sale. However, the seafood industry (including aquaculture) is a $400 billion global industry. The marine capture component accounts for an estimated $212 billion, of which 65 percent, or $140 billion, represents the postharvest economy (Davidsson 2007). The downstream benefits from a more efficient harvest sector are considerable, as illustrated by the examples in box 4.1. The upstream benefits are less evident, though fleet and processing plant construction and modernization can contribute to wealth and economic growth.

The substantial value of noncommercial uses of fisheries is not included in the rent estimates. For example, in the United States, the total national economic impact from commercial finfish fisheries is 28.5 percent of the impact created by marine recreational fisheries (Southwick Associates 2006), and in the case of the striped bass resources, which are shared between the commercial and recreational sectors, anglers harvest 1.28 times more fish, yet produce over 12 times more economic activity as a result (Southwick Associates 2005). Healthy coral reefs provide a further example. In addition to the lost benefits from fisheries, destruction of coral reefs results in an estimated net present loss to society of $0.1 million to $1.0 million per square kilometer of reef (Cesar 1996).

The depletion of global fisheries cannot be attributed solely to fishing. Pollution, destruction of critical habitats (such as wetlands and coastal zones),

The Bering Sea Pollock Conservation Cooperative did not operate under an ITQ system but created the incentives to generate substantial additional rents. This was done by removing the less efficient vessels, extending the fishing season, and allowing the operators to concentrate on product quality. The yield per ton of fish increased by approximately 10 percent, and recovery of by-products such as high-value fish roe increased by 22 percent. The increased benefits occurred in the postcapture operations, but as a result of a more rational harvest regime and investments in the postharvest phase.

The estimated loss of rents in the harvest sector of Peru's anchoveta fishery is on the order of $200 million per year. Fleet capacity is some 2.5 to 3.4 times the capacity required to harvest the total allowable catch set as a function of the maximum sustainable yield, and the capacity of the fish meal plants is some 2.9–3.8 times that required to process the catch. The fishing season in the world's largest fishery has been reduced to less than 60 days per year with substantial loss of quality and wastage. If, under a rationalized and modernized postharvest sector, the current production of lower-grade fish meal graduated to higher-grade fish meal and a greater recovery of fish oil, the additional net postharvest revenues generated would be an the order of $228 million per year.

Source: Wilen and Richardson 2003; Paredes 2008.

invasive species, climate change, and mineral extraction all play a role. However, fishing is considered the greatest single cause of such depletion (Millennium Ecosystem Assessment 2005).

The possible effect of a reduction in discards is not captured in the model. Although by definition, discards generally have no commercial value to the discarder, they may have an economic value. It is likely that under improved fisheries management—a necessary step to gain the full benefits from fisheries— the catch of previously discarded juveniles of commercially valuable species would be reduced. As a consequence, the sustainable yield of valuable species would probably increase, with a further increase in the estimate of potential rents. For example, if reduced discards were to result in an increase of 5 million metric tons in the global MSY, the estimate of forgone rents loss would increase by some $6 billion per year.

4.4.2   The Effects of Higher Fuel and Food Prices

The impact of higher fuel and food prices on the rent estimate is unclear. The effect depends on the interplay between the impact of the fuel price change on the level of fishing effort; the price elasticity of demand for fish in economies where the cost of the entire food basket increases; and the changes in per capita

Table 4. 4	Effects of Fuel and Food Prices Increases on Economic Rents	
Fuel price increases may:		**Food price increases may:**
Increase rents: • if fishing effort decreases as a result of higher costs • if fishing patterns change to less fuel-intensive modes		Increase rents: • if the increase in fish prices more than compensates for higher harvest costs • if forage (fish meal) fisheries redirect catches to higher-value food products
Decrease rents: • if fuel subsidies increase • if the aggregate global fishery becomes less profitable		Decrease rents: • if lower-cost aquaculture products permanently capture market share from marine capture fishery products • if food price increases stimulate increased fishing effort

Source: Authors.

incomes, which underlie the demand for fish. The outcome of this interplay is likely to be specific to the economy of an individual fishery and the markets for the products of that fishery (table 4.4).

Fuel constitutes a significant part of the cost of fishing, and the price of fuel almost doubled between 2004, the base year, and 2007 (U.S. Energy Information Agency 2007). The recent fluctuations in oil prices greatly increase uncertainty, but the gradual increase in oil prices between 2004 and 2008 suggest that the cost of fishing in the base year may underestimate the cost in the future. For example, the variable costs of fishing effort in March 2008 were some 10 percent higher than they were in 2004, given the share of fuel in variable fishing costs. This increase would reduce the estimated rents loss compared with the base year 2004 by about $4 billion.

4.5 SENSITIVITY ANALYSIS AND CONFIDENCE INTERVALS

The rents loss estimates range from a minimum $30 billion (the logistic function and a 10 percent lower MSY) to over $90 billion (the Fox function and 20 percent higher MSY). The results of the sensitivity of the rents loss estimates up to 20 percent deviations in the input data are illustrated in figure 4.1 for the logistic (Schaefer) and the Fox biomass growth functions, respectively.

As can be seen in the figure, the rents loss estimate is most sensitive to changes in the assumed global maximum sustainable yield and in the volume of landings in the base year. When the values for the other input data are kept constant, the estimated rents loss increases with an increase in the value of the MSY estimate and decreases as the value of landings in the base year increases. The estimated rents loss is much less sensitive to changes in the values for other

Figure 4.1   Sensitivity Analysis of the Results for the Logistic and Fox Models

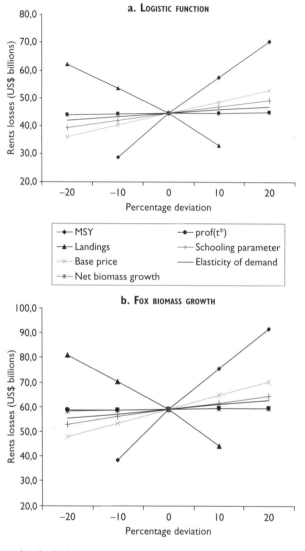

Source: Authors' calculations.

input data such as the price of landed catch, the schooling parameter, and the elasticity of demand.

Based on stipulated stochastic distributions for the input data and calculated stochastic distribution of the rents loss estimates, a 90 percent confidence interval for the estimated rents loss is $31 to $70 billion, with the most probable estimate on the order of $50 billion (table 4.5).

Table 4.5 Confidence Intervals for Rent Loss Estimate

Confidence interval	Range of estimated rents loss (US$ billions)
95 percent	26–73
90 percent	31–70
80 percent	37–67

Source: Authors' calculations.

Figure 4.2 Density and Distribution Functions for the Estimated Rents Loss for Logistic, Fox, and Combined Logistic and Fox Functions

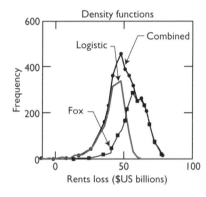
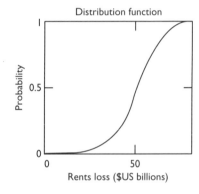

Source: Authors' calculations.

Details of the stochastic distributions for the input data and calculations of the resulting stochastic distribution of the rents loss estimates are described in detail in appendix 3. The stochastic distribution of the rents loss estimates is non-normal and skewed to the right (longer tail to the left). Combining the logistic (Schaefer) and the Fox models in one distribution with equal probability leads to density and distribution functions as illustrated in figure 4.2.

NOTE

1. A representative series of studies using a common methodology is currently being undertaken by FAO and the World Bank under the World Bank's PROFISH Partnership. Results of several of these studies are presented in table A4.4 in appendix 4.

CHAPTER FIVE

The Way Forward

5.1 FISHERIES REFORM MAKES ECONOMIC SENSE

The study shows that an increasing number of fish stocks are overexploited; overcapacity in fishing fleets remains high; the real income level of fishers remains depressed; and fish prices have stagnated, even as the costs of harvesting continue to increase. Aquaculture has grown to approximately 50 percent of food fish production, which has contributed to supply and price stabilization as demand for seafood has increased, particularly in China.

Many thriving and profitable fisheries disguise the fact that at the aggregate level, the economic health of the world's marine capture fisheries is in a state of chronic and advancing malaise that has compromised resilience to fuel price increases, to depressed fish prices, and to the effects of climate variability and change. The estimated loss of potential net benefits is on the order of $50 billion per year for a cumulative loss of over $2 trillion since 1974. The annual loss is equivalent to approximately 64 percent of the landed value of the global catch, or 71 percent of the value of global fish trade in the base year (2004). These estimates, however, exclude the additional value of the environmental benefits of healthy marine ecosystems (such as tourism benefits from healthy coral reefs) and the value of efficiency gains along the value chain. In addition, the full costs of illegal fishing activities and subsidies may not be fully reflected, and the estimated loss of potential benefits is thus conservative.

These are among the many reasons why the economic objectives—increasing the net benefits and wealth from fisheries—need to be at the center stage of efforts to resolve the crisis in marine fisheries. Public awareness and understanding of the potential and actual flows of economic benefits can inform the political economy of reform and help leaders move toward socially responsible and sustainable fisheries underpinned by sound scientific advice. National fisheries policies would benefit from a greater focus on maximizing net benefits and choosing economic or social yield as an objective rather than continuing to manage fisheries with the purely biological objective of maximum sustainable yield as the key reference point.

5.2 REBUILDING GLOBAL FISH CAPITAL

Most marine wild fish resources are considered to be the property of nations. Governments are generally entrusted with the stewardship of these national assets, and their accepted role is to ensure that these assets are used as productively as possible, for both current and future generations. The depletion of a nation's fish stocks constitutes a loss of national wealth, or the nation's stock of natural capital. Similarly, the depletion of global fish stocks constitutes a loss of global natural capital. The scale of these losses— the sunken billions—justifies increased efforts by national economic policy makers to reverse this perennial hemorrhage of national and global economic benefits.

There is enormous potential to rebuild global fish stocks and wealth and to increase the net benefits that countries could derive from their commercial marine fisheries resources. The rents may not be fully recoverable and efforts to rebuild global fish wealth incur economic, social, and political costs. Nevertheless, the sheer scale of the rent drain provides ample grounds for economic policy makers and planners to direct their attention to the rebuilding of national, regional, and global fish capital. Economically healthy marine fisheries can deliver an unending flow of economic benefits, a natural bounty from good stewardship, rather than constituting a net drain on society and on global wealth.

Rising fuel prices, declining fish stocks, and the need for greater fish stock resilience in the face of additional climate change pressures further reinforces the arguments for concerted national and international actions to rebuild fish wealth. Rising food prices, a growing fish food gap for over 1 billion people dependent on fish as their primary source of protein, and the ungainly carbon footprint of some fisheries add to the rationale for reversing the rent drain.

5.2.1 Subsidies

The increase in the prices of fuel and food during 2008 combined to strengthen pressure for subsidies. Such pressures stem not only from the harvest sector of the fisheries industry but also from the upstream and downstream economy dependent on the sector, and from consumers in countries where fish is a staple component of the diet.

The World Bank has recently addressed the subsidies issue. The World Bank does not advocate subsidies as a response to recent food and energy price increases, but supports careful analysis, monitoring, and balancing of competing needs for energy and food security (World Bank 2008a).

The World Development Report 2008 (World Bank 2007c) poses two questions with regard to input subsidies. First, "do the economic benefits exceed the costs of subsidies?" The evidence presented in this and other studies shows that, in the case of fisheries the answer is almost invariably "no" and that the negative environmental externalities generated by input subsidies are considerable.

The second question is "are input subsidies justified on social grounds?" The answer depends on whether the alternatives are more cost-effective. In the case of fisheries, subsidies often constitute a politically expedient means of sidestepping the challenge of addressing the alternatives, including the challenge of helping fisher households to take up other gainful economic opportunities. Often conceived as a short-term intervention, subsidies tend to become entrenched at high cost to society and frequently confer more benefits on the more affluent (for example, vessel owners) than on the intended targets (for example, vessel crew or the poorer consumers). The use of subsidies implies that solutions to the crisis in fisheries lie within the sector rather than through local, regional, and national economic growth. By creating perverse incentives for greater investment and fishing effort in over-stressed fisheries, input subsidies tend to reinforce the sector's poverty trap and undermine the creation of surplus that could be invested in alternatives, including education and health.

The World Bank has suggested, that if input subsidies are to be used, they should be temporary, as part of a broader strategy to improve fisheries management and enhance productivity. The Bank has emphasized investing in quality public goods, such as science, infrastructure, and human capital; improving the investment climate and access to credit; and strengthening governance of natural resources, including through secure user and property rights and collective action taken by a strengthened civil society (World Bank 2008a).

5.2.2 The Costs of Reform

The transition to economically healthy fisheries will require investment. Assessment of the costs of reform and the improved governance required to capture increased net benefits from marine capture fisheries lies beyond the scope of this study, as does an assessment of the proportion of the potential net benefits that can feasibly be captured. The benefits from stock recovery accrue over a longer period and are shrouded in the uncertainties of the ecosystem.

Public funds have been used to finance various elements of reform including fisher retraining and early retirement. Buyback schemes are one of the many strategies deployed to improve the economic performance of fisheries and are generally financed by public funds, although some cost recovery has accrued through charges on the remaining fishers. In Norway, Japan, and elsewhere, private funds have supported buybacks (Curtis and Squires 2007), and dedicated financial instruments have also been proposed (Dalton 2005).

The recurrent costs of management are not addressed in the model presented here. Substantial investment is needed in the transition process to economically healthy fisheries. The investment is required not only in building technical capacity for fisheries management but in the institutional fabric of fisheries tenure at all levels—the fishers, the administration, and the political levels. The recurrent costs of fishery management may decline under an economically healthy fisheries regime. For example, illegal fishing is likely to decline and the costs of enforcement may decline. The cost of the regulatory burden on the fisher may also decline. The allocation of the management cost burden between public and private sectors presents challenges both for fiscal policy and management practice.

5.2.3 Net Benefits and Tenure

It has long been understood that because the benefits of use are individual but costs are shared, the net benefits from use of common pool resources, such as fish stocks, will tend to dissipate (Gordon 1954; Hardin 1968). The nature of the rights over the resources plays an important role in determining the extent of that loss of net benefits, and it is suggested that in general the more clearly defined and enforceable the rights, the less the benefit loss (Scott 1955). In many countries, marine fishery resources are considered to belong to the nation, and governments are charged with the stewardship of this public trust. In some instances, this has undermined the traditional rights systems observed by local communities and led to a de facto open access condition. Because the public or common pool character of marine fish resources is often deeply embedded in law and practice, strengthening marine fisheries is often a complex undertaking that faces political, social, and legal challenges, requiring a good understanding of traditional rights systems, accepted practices, and culture. Nevertheless, to increase the net benefits from fisheries, the issue of tenure must be addressed (de Soto 2000).

The purpose of this study is not to be prescriptive with regard to marine fisheries tenure, but to raise awareness of this link between tenure and net benefits (Costello, Gaines, and Lynham 2008). A greater understanding of this link implies public awareness of the potential and actual economic benefits from marine fisheries and how these benefits can be captured rather than dissipated. It calls for public awareness concerning who benefits and to what extent society underwrites those benefits. It calls for greater understanding of how a balance between secure tenure and the social responsibility for resource stewardship can be achieved at local and national levels. Figure A4.1 in appendix 4 demonstrates that quantifies the increasing wealth generally attributed to strengthened tenure in selected New Zealand and Icelandic fisheries.

5.2.4 Sustainable Fisheries Are Primarily a Governance Issue

As stated in the World Summit on Sustainable Development Plan of Implementation, sound science and an ecosystem approach are fundamental underpinnings of sustainable fisheries (Articles 30, 36). However, the principal drivers of the overexploitation in marine capture fisheries and the causes of the dissipation of the resource rents and loss of potential economic benefits are the perverse economic incentives embedded in the fabric of fisheries harvesting regimes, reflecting a failure of fisheries governance.

Sustainable fisheries are primarily a governance issue, and the application of the fishery science without addressing the political economy of fisheries is unlikely to rebuild marine fish wealth.[1] Restoration of marine fish wealth and rebuilding the flow of net benefits implies fisheries governance reforms with an increased emphasis on the economic and social processes, informed by, rather than centered on, biological considerations and recognizing solutions and opportunities provided in the broader economy outside the fisheries sector.

5.2.5 Fishery Reform Can Advance along the Axes of Sustainability, Productivity, and Equity

Three axes of reform can be considered. A sustainability axis would maintain ecosystem and intergenerational integrity while underpinning the physical basis for economic health. A productivity axis would aim to maximize rents by focusing on the economic efficiency of the harvesting regime. An equity axis would qualify the productivity aspiration, addressing the social dimension of resource allocation or benefit flows.

The maximum economic yield (or a similar proxy) is generally a more conservative harvesting target than maximum sustainable yield (Grafton, Kompass, and Hilborn 2007). Framed within a broader ecosystem approach, it satisfies both the sustainability and rent-maximizing objectives. Advancing along the equity axis, the use of fisheries as a social safety net, for example, may involve some sacrifice of the productivity targets. By contrast, a narrow focus of reform on productivity and rent maximization will fail to address the real social and political costs of rebuilding fish wealth.

A reform agenda calls for a greater understanding of the political and social processes and drivers of change in fisheries. It calls for approaches to dismantling perverse incentives through appropriate tenure and property rights systems and the phasing out of subsidies that enhance fishing effort and fishing capacity. Guidance is available on some elements of reform processes, such as limited entry (Townsend 1990; Cunningham and Bostock 2005); buyback schemes (Curtis and Squires 2007; Clark, Munro, and Sumaila 2007); and individual transferable quotas and property rights (Committee to Review Individual Fishing Quotas 1999; Shotton 1999; WHAT 2000; Grafton et al. 2008). Guidance is also available on community rights (Christy 1999; Willmann 2000); on governance and corruption (World Bank 2007a; World Bank and IUCN in press); and on the political economy of reform and the durability of reforms (OECD 2008; Kjorup 2007). However, greater knowledge is required about the assessment and mitigation of social and political costs, the financing of reform, the timescale and sequencing of reform activities within political and investment cycles, and consensus building among competing stakeholders and their political constituencies. Fisheries reform can also be seen as part of a broader public policy agenda embracing fiscal reforms, pathways out of poverty, and greater transparency in stewardship and accounting for natural capital.

A constructive dialogue on the political economy of reform requires a common understanding among stakeholders of the potential net benefits from marine fisheries and of the current level of benefits and transparency in the allocation of those benefits. A constructive dialogue on reform will require knowledge of the political and social costs and benefits of reform options and informed stakeholder discussion on the alternatives (including transitions out of fisheries). Reforms may take time and require forging a political consensus and vision spanning changes of government. Experience shows that successful reforms may require champions or crises to catalyze the process.

5.2.6 Strengthening the Socioeconomic Dimension of the Fisheries Dialogue

A target set out in the World Summit on Sustainable Development Plan of Implementation is the restoration of fish stocks to maximum sustainable yield levels by 2015. Harvesting at the MSY level is unlikely to capture a substantial part of the economic rents and can be regarded as a minimum target. The MSY target also implies a focus on the fish, and tilts toward a single species approach, rather than focusing on the underlying economic drivers, the political and social challenges to sharing the fish wealth, and the process of reform.

Nevertheless, as a first step in tracking progress toward the restoration of fish stocks, countries, the primary global stakeholders, could report both on the state of fish stocks within their jurisdictional waters (see, for example, NMFS 2008; Department of the Environment, Water, Heritage and the Arts 2008), and on the level and distribution of benefits from the national fish wealth.

5.2.7 Accounting for Fish Wealth Is a National Role

It is a matter of considerable concern that the depletion of fish wealth—natural capital—normally does not show up in the national accounts of countries. Because of weak property rights in national and international fisheries and because of difficulties in establishing market prices for these resources, fisheries assets fall outside the asset boundary of the System of National Accounts 1993 (the internationally recognized system for maintaining national accounts). As a result, a fishing country can run down fish resources and thus temporarily increase catch rates, which show up as an addition in the national accounts, without having to subtract the corresponding reduction in fish stock capital. In other words, fishing nations have drawn upon the fishing sector's opaque natural capital account to "artificially" improve the nation's GDP and simultaneously use this capital to (temporarily) support the operating accounts of fishers and the fishing companies.

Ideally, the system of national accounts should, as a matter of course, include changes in natural capital just as it does for man-made capital. Given their economic importance, the omission of natural assets such as fish stocks from the national accounts entails a substantial oversight in economic accounting. National accounts including changes in natural capital are often referred to as green accounts, and specific guidance is readily available on environmental accounting for fisheries (UN and FAO 2004, Danielsson 2005). Because of the deficit of information on the economic health of the world's fisheries, the World Bank report "Where Is the Wealth of Nations?" was unable to take account of fisheries. Greater awareness of the scale of this capital asset depletion at the level of national policy makers and economic planners could build support for reform processes.

5.2.8 Rights to Harvest Fish Wealth Are Distinct from Rights to Benefit from Fish Wealth

The notion that harvesters (fishers) have an exclusive, rather than a partial and conditional, right to the benefits from marine fisheries has tended to obscure the quest for increased social and economic benefits to society as a whole. This study shows that, in aggregate, the benefits to society as a whole are negative—that society underwrites the sector, through subsidies, by paying the costs of fisheries management, and through depletion of capital (fish wealth).

Rights and obligations are mutually supporting elements of governance, and strengthened marine resource property rights demand both clarity on and respect for the accompanying obligations (Fisman and Miguel 2006).

Many traditional regimes distinguished rights to harvest from rights to benefits in acknowledgement that society at large also had a claim to the benefits of the harvest (Johannes 1978). The same principles are successfully applied in a modern setting, for example in fisheries in New Zealand (see figure A4.1 in appendix 4) and the Shetland Islands, where the tenure is vested in the community and harvest rights are largely "firewalled" from the fundamental wealth creation and capital formation functions.

5.3 SUMMARY: THE WAY FORWARD

1. Use the results of this study to raise **awareness** among leaders, stakeholders, and the public on the potential economic and social benefits from improved fisheries governance.
2. Foster **country-level and fishery-level estimates** of the potential economic and social benefits of fisheries reform and of the social and political costs of reform as a basis for national-, or fishery-level dialogue.

3. Build a portfolio of experiences on the process of fisheries reform with a focus on **the political economy of reform**, process design, change management, social safety nets, and the timescale and financing. Draw on the knowledge and lessons of reforms in other sectors, in particular with regard to the impact on the poor and the effectiveness and equity of adjustment mechanisms.

4. Progressively identify a portfolio of **reform pathways** based on a consensus vision for the future of a fishery founded on transparency in the distribution of benefits and reforms that increase social equity. Common elements of such pathways could include effective stakeholder consultation processes; sound social and economic justifications for change; and an array of social and technical options, including decentralization and comanagement initiatives to create more manageable fishery units. A reform process will bend the trusted tools of fisheries management to new tasks. Sound scientific advice, technical measures such as closed seasons, and effective registration of vessels and existing fishing rights are likely to form synergies with poverty reduction strategies, transitions out of fisheries, social safety nets, and community comanagement.

5. Review fiscal policies in order to phase out **subsidies** that enhance fishing effort and fishing capacity and to redirect public support measures toward strengthening fisheries management capacities and institutions and avoiding social and economic hardships in the fisheries reform process.

6. In an effort to comply with the World Summit on Sustainable Development Plan of Implementation call for restoration of fish stocks, countries could, on a timely basis provide to their public an assessment of the **state of national fish stocks** and take measures to address the underreporting or misreporting of catches.

7. Countries can further justify reforms in fisheries by recognizing that responsible fisheries build resilience to the effects of **climate change** and reduce the carbon footprint of the industry.

NOTE

1. The Code of Conduct for Responsible Fisheries (FAO 1995) provides an overarching framework for sustainable fisheries.

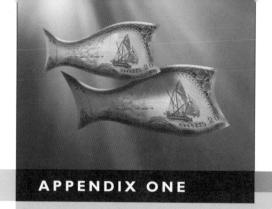

The Concept of Economic Rent in Fisheries

conomic rent is defined as "the payment (imputed or otherwise) to a factor in fixed supply."[1] This definition is formulated in terms of a factor of production and can be extended to cover any restricted variable, such as fish catch.

Figure A1.1, showing a demand curve and a supply curve, is often used to illustrate Ricardo's theory of land rents. In the figure, the market price is p. However, because the quantity of the factor (for example, land) is fixed, the corresponding supply, y, would be forthcoming even if the price were zero, and so the price, p, may be regarded as a surplus per unit of quantity. The total surplus is represented by the rectangle $p \cdot y$, which also represents the economic rents attributable to the limited factor, y.

The economic rents depicted in figure A1.1 represent rental income to the owner of the factor in fixed supply (for example, land) who rents it out to users. The economic rents do not, however, represent the total economic benefits of the supply y. These benefits are measured by the sum of economic rents and the demanders' surplus represented by the upper triangle in the diagram. Thus, in the case depicted in figure A1.1, total benefits, those of the owner plus those of the demanders,[2] would be greater than economic rents.

However, in fisheries (as, indeed, in most other natural resource use), the quantity of supply is not fixed. At each point of time, it is usually possible to extract more or less from the resource stock. Usually in common pool fisheries, the demand will push the supply to y_0 (see figure A1.1), at which point there are no economic rents. At the other extreme, supply may be limited by

Figure A1.1 Economic Rents

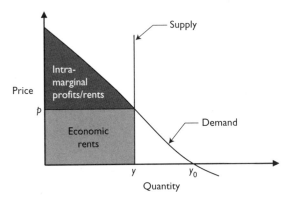

Source: Authors' depiction.

a management regime with the objective of maximizing fisheries rents. Between these extremes, the various fisheries management regimes restrict the harvest quantity at different levels and in different ways.

A cost is associated with resource reduction (a variant of capital reduction cost) for each level of harvest from the stock. This is entirely separate from the cost of the harvesting activity as such, which is included in the demand curve. This cost is the economically appropriate supply price of fish.[3] The resource reduction cost increases with the quantity extracted, or level of harvest. This defines an economically appropriate supply curve for harvest (Arnason 2006), as illustrated in figure A1.2.

The optimally managed fishery will set the actual quantity of supply (allowable harvest) at y, corresponding to the intersection between the supply and demand curves in A1.2. At this level point, there will be a price of supply denoted by p in the diagram. The supply y gives rise to fisheries rents as indicated by the rectangle in the figure.[4] Under conditions of open access, the supply is not restricted and the quantity of extraction will be at y_0 which corresponds to no rents at all.

Measurement of fisheries rents means estimating areas represented by such rectangles and requires estimates of the demand curve for harvests. The demand curve for harvests follows from the profit function of the fishing industry. A simple form of this function is written as

$$\Pi(y, x),\tag{1.1}$$

where y is the harvest level and x the biomass of the stock. The demand curve for harvest is defined as the instantaneous marginal profits from harvest (Arnason 2006) and may be written as

$$\Pi_y(y, x).\tag{1.2}$$

Figure A1.2   Illustrative Resource Rents in a Resource Extraction Industry

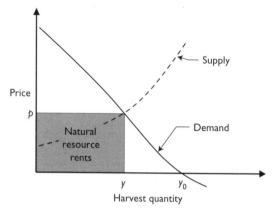

Source: Authors' depiction.

Accordingly, fisheries rents are defined as

$$R(y,x) = \Pi_y(y,x) \cdot y. \qquad (1.3)$$

An estimation of fisheries rents requires a determination of the marginal profits of the fishing industry. To estimate maximum economic rents or economic rents in equilibrium, a bioeconomic model of the fishery is needed.

NOTES

1. As defined by Alchian (1987) in the *New Palgrave Dictionary of Economics* and building on the classical theory by Adam Smith (1776) and David Ricardo (1817).
2. Some authors refer to the demanders' surplus as intramarginal rents. See, for example, Coglan and Pascoe (1999) for the case of fisheries and Blaug (2000) more generally. DFID (2004) provides a short overview of rents in fisheries, and Clark and Munro (1975) provide an overview of fisheries and capital theory.
3. Also called user cost by Scott (1955) and shadow price of the resource by Dasgupta and Heal (1979).
4. The rectangle, represented by the multiple $p \cdot y$ in the figure, corresponds to economic rents in the traditional (Smith-Ricardian) sense as defined by Alchian.

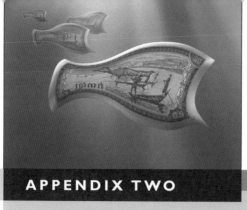

Model and Model Estimation

This appendix sets out the details of the global fisheries model employed in this study and explains how it is applied.

THE BASIC MODEL

The basic model is the following aggregative fisheries model:

$$\dot{x} = G(x) - y \qquad \text{(biomass growth function).} \qquad (2.1)$$

$$y = Y(e, x) \qquad \text{(harvesting function).} \qquad (2.2)$$

$$\pi = p \cdot Y(e, x) \qquad \text{(profit function).} \qquad (2.3)$$

$$R \equiv \Pi_y(y, x) \cdot y = \left(p - C_e(e) \cdot \frac{\partial e}{\partial y} \right) \cdot y \qquad \text{(fisheries rents).}[1] \qquad (2.4)$$

Equation 2.1 describes net biomass growth, denoted by the derivative, $\dot{x} \equiv \partial x / \partial t$. The variable x represents the level of biomass and y harvest. The function $G(x)$ represents the natural growth of the biomass before harvesting. Equation 2.2 explains the harvest as a function of fishing effort, e, and biomass. Equation 2.3 defines profits as the difference between revenues, $p \cdot Y(e, x)$, where p denotes the average net landed price of fish and costs are represented by the cost function $C(e)$. Equation 2.4 specifies fisheries rents, R. This, as explained in appendix 1, is formally defined as $(\partial \pi / \partial y) \cdot y$.

Of the six variables in this model, that is, x, y, π, R, p, and e, the first four may be seen as endogenous, that is, determined within the fishery. The fifth, price, is exogenous, determined by market conditions outside the fishery. The sixth, fishing effort, e, may be seen as the control variable, that is, the variable whose values may be selected to maximize benefits from the fishery.

THE SPECIFIC MODEL

The basic model comprises three elementary functions; the natural growth function, $G(x)$; the harvesting function, $Y(e, x)$; and the cost function, $C(e)$. The specific model is defined by deciding on the form of these functions.

Two variants of the biomass growth function $G(x)$ are used: the logistic function (Volterra 1923) and the Fox function (Fox 1970). As explained previously, the main difference between these two functions is that the Fox function exhibits higher biomass growth at relatively low biomass levels and thus is more resilient to high levels of fishing effort than the logistic function.

$$G(x) = \alpha \cdot x - \beta \cdot x^2, \qquad \text{(Logistic)} \qquad (2.5)$$

$$G(x) = \alpha \cdot x - \beta \cdot \ln(x) \cdot x \qquad \text{(Fox 1970).} \qquad (2.6)$$

For harvesting, the generalized Schaefer (1954) form is selected:

$$Y(e, x) = q \cdot e \cdot x^b, \qquad (2.7)$$

where the coefficient b indicates the degree of schooling behavior by the fish (normally $b \in [0,1]$). The coefficient q is often referred to as the catchability coefficient.

For the cost function, the following linear form is chosen:

$$C(e) = c \cdot e + fk, \qquad (2.8)$$

where c represents marginal variable costs and fk fixed costs.
Under these functional specifications the complete model becomes:

$$\dot{x} = \alpha \cdot x - \beta \cdot x^2 - y, ,$$

or   (biomass growth functions).   (2.9)

$$\dot{x} = \alpha \cdot x - \beta \cdot \ln(x) \cdot x - y$$

$$y = q \cdot e \cdot x^b \qquad \text{(harvesting function).} \qquad (2.10)$$

$$\pi = p \cdot y - c \cdot e - fk \qquad \text{(profit function).} \qquad (2.11)$$

$$R = p \cdot y - \left(\frac{c}{q}\right) \cdot y \cdot x^{-b} \qquad \text{(fisheries rents).} \qquad (2.12)$$

Assuming biomass equilibrium, that is, $\dot{x} = 0$, it is possible to deduce from equations 2.9 and 2.10 the equilibrium or sustainable yield curves as functions of fishing effort for the two biomass growth functions. The corresponding equilibrium revenue curves are illustrated in figure A2.1, where the graph of the cost curve is also depicted. The resulting equilibrium diagram is usually referred to as the sustainable fisheries model (see, for example, Hannesson 1993).

The discontinuity in both equilibrium revenue functions illustrated in figure A2.1 is a common feature in real fisheries (see, for example, Clark 1976). In this particular case it occurs because a degree of schooling behavior ($b < 1$) has been assumed.

Equilibrium profits from the fishery are maximized at a fishing effort level where the distance between equilibrium revenues and costs is greatest. As can be seen from figure A2.1, this occurs at different fishing effort levels for the two biomass growth functions.

Equilibrium fisheries rents are not generally identifiable from a diagram such as figure A2.1, and fisheries rents are generally not maximized at the same effort level that maximizes profits. However, for the specific model of this study, rents may be identified as the difference between equilibrium revenues and the variable costs curve (that is, a curve parallel to the cost curve but passing through the origin). Also, in this specific model, the rents and profits maximizing fishing effort levels coincide, although maximum rents may well exceed maximum profits.

A condensed form of the model may be obtained by combining equations 2.10 and 2.11 to yield

$$\pi = p \cdot y - \left(\frac{c}{q} \right) \cdot y \cdot x^{-b} - fk \quad \text{(profit function)}. \tag{2.13}$$

This condensed form of the model, that is, equations 2.9, 2.12, and 2.13, shows that knowledge of fishing effort is not needed to run the model, and that

Figure A2.1   The Equilibrium Fisheries Model

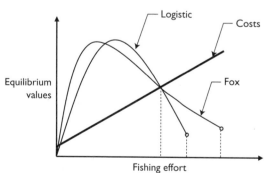

Source: Authors' depiction.

marginal costs and catchability, c and q, do not play an independent role in this model. The ratio of the two (c/q) may be regarded as a single coefficient, referred to as "normalized marginal cost."

ESTIMATION OF MODEL INPUTS

The "specific fisheries model" (that is, equations 2.9, 2.12, and 2.13) contains six unknown coefficients $\alpha, \beta, \left(\dfrac{c}{q}\right), b, p, fk$. These have to be estimated from data or determined in some other way. The model also contains five unknown variables, namely, x, y, π, and R as well as the change in biomass, $\dot{x}$. The model can be used to solve for three of these variables endogenously. The other two have to be either estimated from data or determined in some other way. In the rents loss calculations of this study, current or base year rents are compared to maximum equilibrium rents. For the calculations of maximum equilibrium rents, estimates of these two variables are not required. First, the equilibrium biomass is constant, so $\dot{x} = 0$. Second, the harvest, y, is determined by the maximization exercise. For the current rents calculations, estimates of base year harvest and biomass growth, $y(t*)$ and $\dot{x}(t^*)$, respectively, were obtained. The model inputs (coefficients and variables) that have to be estimated are listed in table A2.1.

There are many ways to obtain estimates of the model input data listed in table A2.1. Because the quality of some global fisheries data sets is poor, the study has elected a procedure that minimizes data requirements. The global data needed are listed in table A2.2. The procedure is summarized as a series of estimation formulas listed in table A2.3. These formulas can be verified by the appropriate manipulation of the specific model above.

The change in fishing effort from an initial to an optimal fishery can be calculated using the same basic data listed in table A2.2. More precisely, it can be shown that

$$e* = \dot{\varphi} \cdot e(t*), \quad \text{where } \varphi = \frac{p \cdot y* - \Pi *}{p \cdot y(t*) - \Pi(t*)}, \text{ and} \qquad (2.14)$$

where '*' indicate the final equilibrium levels of variables and '$t*$' the base year values.

If the numerical value of $e(t*)$ is known, the numerical value of $e*$ can be calculated for this expression. Otherwise, $e*$ can be calculated as a fraction of $e(t*)$, that is, as an index.

The input values used for the estimations and their respective sources are listed in table A2.4.

In a long-run economic equilibrium, all costs are variable (Varian 1984). This is because in the long-run equilibrium, all capital (the source of fixed costs) has been adjusted. Therefore, in the movement to long-run equilibrium, all so-called fixed costs are in fact variable. In this study, the equilibrium (or

Table A2.1 Summary of Model Coefficients and Variables That Need to Be Estimated

Coefficient or variable		Characterization	Permissible values
Biological coefficients			
Biomass growth function	α	Intrinsic growth rate (only for the logistic function)	$\alpha > 0$
Biomass growth function	β		$\beta > 0$
Harvesting function	b	Schooling parameter	$0 < b \leq 1$
Economic coefficients			
Cost function	c/q	Marginal cost ratio	$c/q > 0$
Cost function	fk	Fixed costs	$fk \geq 0$
Revenues	p	Net landings price	$p > 0$
Variables (in base year, t*)			
Landings	$y(t^*)$	Volume of landings	$y(t^*) \geq 0$
Biomass growth	$\dot{x}(t^*)$	Biomass growth	

Source: Authors' models.

Table A2.2 Data for Estimation of Model Coefficients and Variables

Biological data	Symbol
Maximum sustainable yield	MSY
Biomass carrying capacity	X_{max}
The schooling parameter	B
Fisheries data in a base year t*	
Biomass growth in year t^*	$\dot{x}(t^*)$
Landings in year t^*	$y(t^*)$
Price of landings in year t^*	$p(t^*)$
Profits in year t^*	$\pi(t^*)$
Fixed cost ratio in year t^* ($fk / TC\,(t^*)$)	$\varepsilon(t^*)$

Source: Authors' models.

long-run) maximum rents are going to be compared to current rents. There-fore, within the framework of this study, any fixed costs experienced in the base year are taken to be variable when considering the movement to the rents-maximizing equilibrium. This is equivalent to setting the fixed cost ratio in the

Table A2.3 Formulas to Calculate Model Parameters

Unknowns	Formula
Logistic function	
$\hat{\alpha}$	$\hat{\alpha} = 4 \cdot \dfrac{MSY}{X_{max}}$
$\hat{\beta}$	$\hat{\beta} = 4 \cdot \dfrac{MSY}{X_{max}^2}$
Biomass in base year, $\hat{x}(t^*)$	$\hat{x}(t^*) = \dfrac{\hat{\alpha}}{2\hat{\beta}} \cdot \left(1 \pm \left(1 - \dfrac{4 \cdot \hat{\beta} \cdot (y(t^*) + x(t^*))}{\hat{\alpha}^2} \right)^{0.5} \right)$
Fox function	
$\hat{\alpha}$	$\hat{\alpha} = MSY \cdot \ln(X_{max}) \cdot \dfrac{exp}{X_{max}}$
$\hat{\beta}$	$\hat{\beta} = MSY \cdot \dfrac{exp}{X_{max}}$
Biomass in base year, $\hat{x}(t^*)$	$\left(\hat{\alpha} - \hat{\beta} \cdot \ln(\hat{x}(t^*)) \right) \cdot \hat{x}(t^*) = x(t^*) + y(t^*)$
Normalized marginal cost, $\left(\dfrac{\hat{c}}{q} \right)$	$\hat{c} = \dfrac{(p(t^*) \cdot y(t^*) - \pi(t^*)) \cdot (1 - \varepsilon)}{y(t^*) \cdot \hat{x}(t^*)}$
Fixed costs, $\hat{fk}$	$\hat{fk} = (p(t^*) \cdot y(t^*) - \pi(t^*)) \cdot \varepsilon(t^*)$
The schooling parameter, $\hat{b}$	B
Landings in year t^*, $\hat{y}(t^*)$	$y(t^*)$
Price of landings in year t^*, $\hat{y}(t^*)$	$p(t^*)$

Source: Authors' models.

base year equal to zero. Note that this does not imply that the fixed costs in the base year are ignored. They are included but regarded as variable costs.

On the basis of the empirical assumptions listed in table A2.4 and the formulas in table A2.3, the model coefficients can be calculated. The results are listed in table A2.5.

With the empirical assumption and the estimates above, the condensed form of the global fisheries model employed in this study becomes:

$$\dot{x} = 0.839 \cdot x - 0.002 \cdot x^2, \qquad \text{(logistic biomass growth).}$$

$$\dot{x} = 3.486 \cdot x - 0.57 \cdot \ln(x) \cdot x, \qquad \text{(Fox biomass growth).} \tag{2.15}$$

$$\Pi(y,x) = 0.918 \cdot y - 32.8 \cdot y \cdot x^{0.7}, \qquad \text{(profits for the logistic).}$$

$$\Pi(y,x) = 0.918 \cdot y - 23.2 \cdot y \cdot x^{0.7}, \qquad \text{(profits for the Fox).} \tag{2.16}$$

Table A2.4 Empirical Values and Assumptions for Estimation of Model Coefficients

Input data		Units	Value
Biological data			
Maximum sustainable yield	MSY	m. metric tons	95
Carrying capacity	X_{max}	m. metric tons	453.0
Fisheries data in base year (2004)			
Biomass growth in base year t^*	$\dot{x}(t^*)$	m. metric tons	−2
Landings in base year t^*	$y(t^*)$	m. metric ton	85.7
Price of landings in base year t^*	$p(t^*)$	1000 $/ton	0.92
Profits in base year t^*	$\Pi(t^*)$	billion $	−5
Fixed cost ratio in base year t^*	$\varepsilon(t^*)$	ratio	0
The schooling parameter	b	no units	0.7
Elasticity of demand with respect to biomass	d	no units	0.24
Effort (index or real base year effort)			
Fishing effort (fleet) in base year	$e(t^*)$	index	1.00

Source: Authors' models.

Table A2.5 Calculated Model Coefficients (Implied)

Variable	Logistic	Fox
Biomass growth parameter, α	0.839	3.486
Biomass growth parameter, β	0.002	0.570
Biomass, x (2004)	148.4	92.3
Normalized marginal costs, c/q	32.3	23.2
Schooling parameter, b	0.7	0.7
Fixed costs, fk	0	0

Source: Authors' calculations.

$$R(y, x) = 0.918 \cdot y - 32.8 \cdot y \cdot x^{0.7} \quad \text{(fisheries rents for the logistic).}$$

$$R(y, x) = 0.918 \cdot y - 23.2 \cdot y \cdot x^{0.7} \quad \text{(fisheries rents for the Fox).}$$

(2.17)

In the same way as in figure A2.1, the essence of the empirical global fisheries model can be illustrated graphically (figure A2.2).

Figure A2.2   Graphical Illustration of the Global Fishery

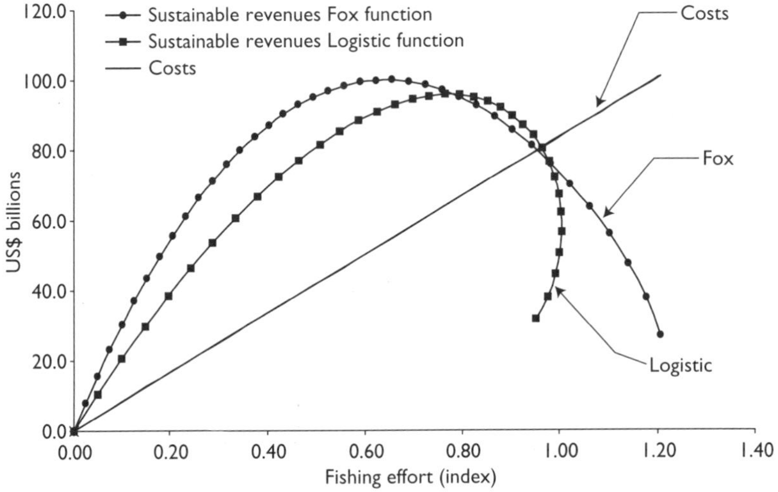

Source: Authors.

NOTE

1. In appendix 1, rent was defined as $R \equiv \Pi_y(y,x) \cdot y$, where $\Pi_y(y,x)$ is the first derivative of the profit function, that is, marginal profits. For this particular fisheries model with fishing effort rather than harvest as a control variable, $\Pi_y(y,x) = p - C_e(e) \cdot \partial e / \partial y$.

Stochastic Specifications
and Confidence Intervals

Because of the uncertainties concerning the empirical values and assumptions underlying the global fisheries model, the resulting rent loss estimates should be regarded as stochastic with associated probability distributions. Because the rents loss calculations involve a complex nonlinear function of the empirical data and assumptions, the analytic equations for the probability distribution of these estimates are not readily obtainable. To generate confidence intervals for the rents loss, reasonable probability distributions for the empirical data and assumptions are specified, and Monte Carlo stochastic simulations (Davidson and MacKinnon 1993; Fishman 1996) were used to generate probability distributions for the model inputs and outcomes (fisheries rents and fisheries rents loss).

Probability distributions for the stochastic input parameters listed below were generated on the basis of 2,000 simulations drawing from the distributions specified here. The stochastic specifications are summarized in table A3.1, and the resulting outcomes and distributions are illustrated in the figures that follow.

1. Global maximum sustainable yield (MSY)
2. Global biomass carrying capacity (X_{max})
3. Biomass growth in the base year ($\dot{x}$)
4. Landings in the base year (y)
5. Landings price (p)
6. Profits in the base year ($prof/\Pi$)
7. Schooling parameter (b)
8. Elasticity of demand (D)

The remaining input parameter, the fixed cost ratio (ε), is assumed to be nonstochastic (see appendix 2).

Table A3.1 Empirical Values and Assumptions: Stochastic Specifications

Variable	Point estimate	Type of distribution	Standard deviation[a]	Implied 95 percent confidence interval
MSY	95	Log-normal	0.03	89.5 to 100.9
X_{max}	453	Log-normal	0.1	370.9 to 553.3
$\dot{x}(t^*)$	−2	Normal	3.0	−8 to 4
$y(t^*)$	85.7	Log-normal	0.015	83.2 to 88.3
$p(t^*)$	0.918	Log-normal	0.03	0.865 to 0.975
$\Pi(t^*)$	−5	Normal	2.5	−10 to 0
b	0.7	Log-normal	0.05	0.63 to 0.77
D	0.2	Log-normal	0.1	0.164 to 0.244
$\varepsilon(t^*)$	0	Log-normal	0.0	0

Source: Authors' calculations.

a. For lognormal distributions, the standard deviation may be interpreted as an approximate percentage deviation.

OUTCOMES OF THE MONTE CARLO STOCHASTIC SIMULATIONS

Logistic Model (figure A3.1)

- Nonnormal distribution
- Mean rents loss: $43.0 billion
- Median rents loss: $44.5 billion
- Mode rents loss (approximately) $48 billion
- Standard deviation: $8.8 billion
- 95 percent confidence interval ($ billion): [20.2, 55.7].

Fox Model (figure A3.2)

- Approximately normal distribution
- Mean rents loss: $59.0 billion
- Median rents loss: $59.2 billion
- Mode rents loss (approximately) $52 billion
- Standard deviation: $9.0 billion
- 95 percent confidence interval ($ billion): [38.8,74.6]

Figure A3.1   Graphical Illustration of Stochastic Simulations Using
the Logistic Model

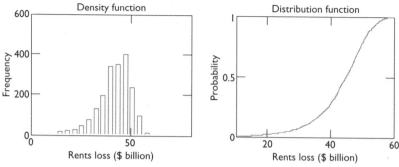

Source: Authors' simulation.

The Combined Models

The difference between the means of the two models appears to be highly significant at the 5 percent level. Assuming that the two biomass growth functions are equally likely, the respective stochastic distributions may be combined in one overall distribution as illustrated in figure A3.3. The crucial outcomes from this combined model are:

- Mean rents loss: $51.0 billion
- Standard deviation: $12.0 billion
- 95 percent confidence interval ($ billion): [26.3, 72.8]

Calculated Rents and Rents Loss

Two thousand draws from the stochastic distributions described earlier were taken and the resulting rents and rents loss calculated. The latter is defined as the difference between the maximum attainable sustainable rents and those that pertain to the base year (2004). Both the current and the maximum rents estimates are stochastic. On this basis, the distributions for the outcomes are derived and confidence intervals calculated. The stochastic specifications for the empirical assumptions are those listed in table A3.1.

The key results of the 2,000 draws from the stochastic distributions described earlier and the resulting rents are summarized in table A3.2 (they are also shown in table 4.5 in the main text). The distribution of the rents loss is illustrated in figure A3.4.

Figure A3.2   Graphical Illustration of Fox Model Stochastic Simulations

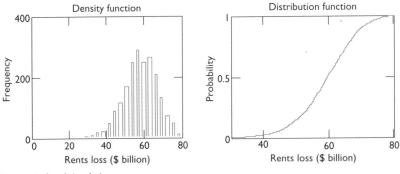

Source: Authors' simulation.

Figure A3.3   Graphical Illustration of Combined Logistic and Fox Model
Stochastic Simulations

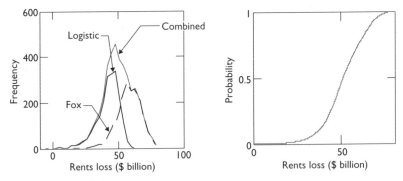

Source: Authors' simulation.

Table A3.2   Estimated Rent Loss: Main Results
(*US$ billions*)

Model	Current (2004)		Maximum sustainable rents		Rents loss	
	Mean	95 percent confidence interval	Mean	95 percent confidence interval	Mean	95 percent confidence interval
Logistic	−5.0	[−10.2, 0.0]	37.6	[4.7, 48.2]	43.0	[20.2, 55.7]
Fox	−5.0	[−10.2, 0.0]	53.4	[41.4, 65.4]	59.0	[38.8, 74.6]
Combined	−5.0	[−10.2, 0.0]	45.3	[38.1, 63.9]	51.0	[26.3, 72.8]

Source: Authors' calculations.

Conclusion

In conclusion, the most reasonable estimate of the global rents loss is:

- Mean: $51 billion per year
  with
- 95 percent confidence interval ($billion per year): [26.3,72.8]
- 90 percent confidence interval ($billion per year): [31.3,69.8]
- 80 percent confidence interval ($billion per year): [36.5,66.9]

DETAILS OF THE PROBABILITY DISTRIBUTIONS FOR THE INPUT PARAMETERS

Maximum Sustainable Yield (MSY)

$MSY^\circ = MSY \cdot e^{u_1}$, $u_1 \sim N(0,\sigma_1)$; where MSY° represents the stochastic maximum sustainable yield and MSY the point estimate. The random term u_1 is assumed to be normally distributed with mean zero and standard deviation σ_1. This specification implies that MSY° exhibits a lognormal distribution. In the stochastic simulations it is assumed $\sigma_1 = 0.03$. This gives rise to the distribution illustrated in figure A3.4(a). An estimated 5 percent confidence interval for MSY° is $MSY^\circ \in [89.5, 100.9]$ million metric tons.

Biomass Carrying Capacity (XMAX)

$XMAX^\circ = XMAX \cdot e^{u_2}$, $u_2 \sim N(0,\sigma_2)$, where $XMAX^\circ$ represents the stochastic carrying capacity of the global commercial biomass with $XMAX$ as the point estimate. The random variable u_2 is assumed to be normally distributed with mean zero and standard deviation σ_2. This specification implies that $XMAX^\circ$ exhibits a lognormal distribution. In the stochastic simulations, it is assumed that $\sigma_2 = 0.1$. This leads to the distribution illustrated in figure A3.4(b). An estimated 5 percent confidence interval for $XMAX^\circ$ is $XMAX^\circ \in [370.9, 553.3]$ million metric tons.

Biomass Growth in Base Year (XDOT)

$XDOT^\circ = XDOT + u_3$, where $XDOT^\circ$ represents the stochastic biomass growth in the base year, and $XDOT$ is the point estimate. The random variable u_3 is assumed to be normally distributed with mean zero and standard deviation σ_3. This specification implies that $XDOT^\circ$ exhibits a normal distribution. In the stochastic simulations, it is assumed that $\sigma_3 = 3$. This generates the distribution illustrated in figure A3.4(c). An approximate 5 percent confidence interval for $XDOT^\circ$ is $XDOT^\circ \in [-8,4]$ million metric tons.

Figure A3.4(a)–(h)   Simulated Distributions

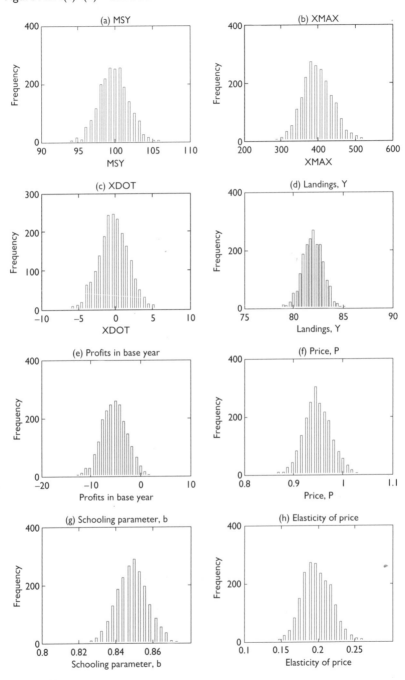

Source: Authors' simulation.

Landings in Base Year (Y)

$Y^\circ = Y \cdot e^{u_2}$, $u_4 \sim N(0,\sigma_4)$, where Y° represents the stochastic landings in the base year and Y is the point estimate. The random variable u_4 is assumed to be normally distributed with mean zero and standard deviation σ_4. This specification implies that Y° exhibits a lognormal distribution. In the stochastic simulations, $\sigma_4 = 0.015$. This gives rise to the distribution illustrated in figure A3.4(d). A 5 percent confidence interval for Y° is $Y^\circ \in [83.2, 88.3]$ million metric tons.

Profits in Base Year (Prof Π)

$PROF^\circ = PROF + u_5$, $u_5 \sim N(0,\sigma_5)$, where $PROF^\circ$ represents the stochastic profits in the base year and $PROF$ the point estimate for these profits. The random variable u_5 is normally distributed with mean zero and standard deviation σ_5. This specification implies that $PROF^\circ$ exhibits a normal distribution. In the stochastic simulations, $\sigma_5 = 2.5$. This leads to the distribution illustrated in figure A3.4(e). A 5 percent confidence interval for $PROF^\circ$ is $PROF^\circ \in [-10, 0]$ billion \$.

Landings Price (P)

$P^\circ = P \cdot e^{u_6}$, $u_6 \sim N(0,\sigma_6)$, where P° represents the stochastic landings price P, the point estimate of the landings price, and u_6 is assumed to be a normally distributed random variable with mean zero and standard deviation σ_6. This specification implies that P° exhibits a lognormal distribution. In the stochastic simulations, $\sigma_6 = 0.03$. This gives rise to the distribution illustrated in figure A3.4(f). A 5 percent confidence interval for P° is $P^\circ \in [0.865, 0.975]$ dollars per kilogram.

Schooling parameter (b)

$b^\circ = b \cdot e^{u_2}$, $u_7 \sim N(0,\sigma_7)$, where b° represents the stochastic schooling parameter with b being the point estimate. The random variable u_7 is assumed to be normally distributed with mean zero and standard deviation σ_7. This specification implies that b° exhibits a lognormal distribution. In the stochastic simulations, it is assumed that $\sigma_7 = 0.05$. This gives rise to the distribution illustrated in figure A3.4(g). A 5 percent confidence interval for b° is $b^\circ \in [0.63, 0.77]$.

Elasticity of Demand (d)

$d^\circ = d \cdot e^{u_2}$, $u_8 \sim N(0,\sigma_8)$, where d° represents the stochastic schooling parameter with d being the point estimate. The random variable u_8 is assumed to be normally distributed with mean zero and standard deviation σ_8. This specification implies that d° exhibits a lognormal distribution. In the stochastic simulations, it is assumed that $\sigma_8 = 0.1$. This gives rise to the distribution illustrated in figure A3.4(h). A 5 percent confidence interval for d° is $d^\circ \in [0.164, 0.244]$.

APPENDIX FOUR

Supplementary Data

Table A4.1   Motorized Fishing Fleets in Selected Major Fishing Countries, 2004

Country/ Economy	Reported global marine catch* (%)	Number of vessels	Tonnage (GT)	Power (kW)
China	17	509,717	7,115,194	15,506,720
EU-15	6	85,480	1,882,597	6,941,077
Iceland	2	939	187,079	462,785
Japan	5	313,870	1,304,000	—
Norway	3	8,184	394,846	1,328,945
Republic of Korea	2	87,203	721,398	16,743,102
Russian Federation	3	2,458	1,939,734	2,111,332

Header spanning "Fishing fleet parameters, 2004" covers Number of vessels, Tonnage (GT), Power (kW).

Source: China: FAO fishery statistical inquiry; EU-15: Eurostat; Iceland: Statistics Iceland (http://www.statice.is); Japan: *Japan Statistical Yearbook 2006* (http://www. stat. go.jp/english/data/nenkan/index.htm); Republic of Korea: *Korea Statistical Yearbook 2005* Vol. 52; Norway: Statistics Norway (http://www.ssb.no) and Eurostat; Russian Federation: FAO fishery statistical inquiry, FAO FishStat; Concerted Action 2004.
Note: Some vessels may not be measured according to the 1969 International Convention on Tonnage Measurement of Ships. The Icelandic data exclude undecked vessels. The Japanese data refer to registered fishing vessels operating in marine waters. The Russian Federation data refer to powered decked vessels with a national license.
*excluding aquatic plants.

Table A4.2   Selected Examples of Relationship between Estimated MSY and Biomass Carrying Capacity

Fishery	MSY (tons)	Carrying capacity (tons)	Multiple (biomass/MSY)
Denmark cod	216	1,443	6.68
Norway cod	602	2,473	4.11
Iceland cod	332	1,988	5.99
Denmark herring	666	4,896	7.35
Norway capelin	2,219	8,293	3.74
Iceland capelin	1,010	3,669	3.63
Bangladesh Hilsa	286	1,084	3.79

Source: International Council for the Exploration of the Sea; FAO.

Table A4.3 Estimation of the Weighted Average Global Schooling Parameter

Species group	Imputed MSY (tons)	Schooling parameter	Weighted
Salmons, trouts, smelts	1,016,854	1.00	0.007
Shads	426,754	0.50	0.002
Miscellaneous diadromous fishes	80,134	0.70	0.001
Flounders, halibuts, soles	1,392,052	1.00	0.014
Cods, hakes, haddocks	13,788,742	1.00	0.137
Miscellaneous coastal fishes	6,935,300	1.00	0.069
Miscellaneous demersal fishes	3,162,243	1.00	0.031
Herrings, sardines, anchovies	25,908,711	0.30	0.077
Tunas, bonitos, billfishes	6,243,122	0.60	0.037
Miscellaneous pelagic fishes	14,322,640	0.50	0.071
Sharks, rays, chimaeras	880,785	0.90	0.008
Marine fishes not identified	10,738,831	0.85	0.090
Crabs, sea spiders	1,333,282	0.70	0.009
Lobsters, spiny rock lobsters	233,825	0.70	0.002
King crabs, squatlobsters	163,513	0.70	0.001
Shrimps, prawns	3,478,304	0.80	0.028
Krill, planktonic crustaceans	528,335	0.50	0.003
Miscellaneous marine crustaceans	1,427,312	0.70	0.010
Abalones, winkles, conchs	139,964	1.00	0.001
Oysters	302,526	1.00	0.003
Mussels	317,852	1.00	0.003
Scallops, pectens	804,349	1.00	0.008
Clams, cockles, arkshells	1,129,231	1.00	0.011
Squids, cuttlefishes, octopuses	3,892,145	0.70	0.027
Miscellaneous marine molluscs	1,596,036	0.90	0.014
Sea squirts and other tunicates	21,331	1.00	0.000
Horseshoe crabs and other arachnoids	3,252	1.00	0.000
Sea urchins and other echinoderms	140,461	1.00	0.001
Miscellaneous aquatic invertebrates	539,994	0.90	0.005
TOTAL	100,965,809	n.a.	0.670

Source: MSY values are the historical maximum catches by species group as reported by Fishstat Plus. The schooling parameters are assumed based on information on schooling parameters for several indicative species.

n.a. = Not applicable.

Table A4.4 Indicative Results of Selected Case Studies on Economic Rents in Fisheries

Fishery	Rent/revenue loss as % of base revenues/landed values			Source
	Base year	Percent	Rent or proxy	
Vietnam Gulf of Tonkin demersal multigear, multispecies	2006	29	rent	Nguyen and Nguyen 2008
Icelandic cod demersal multigear, multispecies	2005	55	rent	Arnason pers. comm.
Namibian hake trawl	2002	136	rent	Sumaila and Marsden 2007
Peruvian anchoveta purse seine[a]	2006	29	rent	Paredes 2008
Bangladesh hilsa artisanal multigear	2005	58	rents	
Gulf of Thailand demersal multigear multispecies	1997	42	net revenues	Willmann et al. 2003
Yemen lobster, artisanal	2008	1,653	net revenues	Shotton pers. comm.
British Columbia salmon fishery	1982	76	rents	Dupont 1990
Cyprus fisheries	1984	5	revenue increase	Hannesson 1986
Small pelagic fisheries in northwest Peninsular Malaysia	1980–90	79	revenue	Tai and Heaps 1996
U.S. Atlantic sea scallop	1995	75		Repetto 2002

U.S. fisheries	2003	net present value	192	Sumaila and Suatoni 2006
New England groundfish	1989	rents	188	Edwards and Murawski 1993
Gulf of Mexico shrimp	1990s	present value	50	Ward 2006
Western and Central Pacific tuna	1996	profit	59	Bertignac et al. 2001
Norwegian trawl	1998	rents	439	Ache et al. 2003
Japan coastal squid	2004	rents	77	Hoshino and Matsuda
Japan Pacific saury stick-held dip-net fishery	2004	rents	89	2007
Lake Victoria Nile perch (freshwater)	2006	rents	61	Warui 2008
Danish mussel	2001–03	landed value	9	Nielsen et al. 2007
Swedish pelagic fishery	2001–03	landed value	50	Nielsen et al. 2007
Faroese pair trawl	2001–03	landed value	19	Nielsen et al. 2007
Norwegian coastal (ITQ)	2001–03	landed value	40	Nielsen et al. 2007

Note: Values presented refer to different economic indicators and are not necessarily comparable. The table is provided to illustrate the fact that in many fisheries, substantial additional net benefits can be derived through responsible fisheries management with a focus on economic and social benefits.

a. Economic returns from pelagic fisheries are highly variable and can be heavily influenced by environmental factors or export markets, not merely by the effectiveness of the management regime.

Table A4.5 Projection of Rent Loss, 1974–2007 ($ billion)

| Year | Global fish stock exploitation status fully + over + depleted | | Rents loss (US$ billion) | Deflator[b] | | Deflated rents loss by year (US$ billions) | Cumulative rents loss at 3.5 percent (US$ billions) |
	Percent	Index		Base	2004		
1974	0.61	0.80	40.5	53.5	0.36	14.8	15
1975[a]	—	—	40.5	58.4	0.40	16.1	33
1976	—	—	40.5	61.1	0.42	16.9	51
1977	—	—	40.5	64.9	0.44	17.9	72
1978	0.59	0.77	39.5	69.9	0.48	18.8	93
1979	0.63	0.82	42.0	78.7	0.54	22.5	120
1980	—	—	42.0	89.8	0.61	25.7	151
1981	0.63	0.82	41.8	98	0.67	27.9	185
1982	—	—	41.8	100	0.68	28.5	221
1983	0.69	0.91	46.3	101.3	0.69	31.9	262
1984	—	—	46.3	103.7	0.71	32.7	305
1985	0.68	0.90	45.7	103.2	0.70	32.1	349
1986	—	—	45.7	100.2	0.68	31.2	393
1987	0.69	0.90	45.8	102.8	0.70	32.1	440
1988	—	—	45.8	106.9	0.73	33.3	490
1989	0.69	0.91	46.4	112.2	0.76	35.5	544
1990	0.69	0.90	45.8	116.3	0.79	36.3	601
1991	—	—	45.8	116.5	0.79	36.4	659
1992	0.71	0.93	47.2	117.2	0.80	37.7	721

1993	—	—	47.2	118.9	0.81	38.2	786
1994	—	—	47.2	120.4	0.82	38.7	854
1995	0.70	0.92	47.0	124.7	0.85	39.9	925
1996	—	—	47.0	127.7	0.87	40.9	1,000
1997	0.73	0.96	49.0	127.6	0.87	42.6	1,079
1998	—	—	49.0	124.4	0.85	41.5	1,159
1999	—	—	49.0	125.5	0.86	41.9	1,243
2000	0.75	0.98	48.8	132.7	0.90	44.1	1,333
2001	—	—	48.8	134.2	0.91	44.6	1,425
2002	—	—	48.8	131.1	0.89	43.6	1,520
2002	—	—	48.8	131.1	0.89	43.6	1,520
2003	—	—	48.8	138.1	0.94	45.9	1,621
2004	0.76	*1.00*	51.0	146.7	*1.00*	51.0	1,731
2005	—	—	51.0	157.4	1.07	54.7	1,848
2006	—	—	51.0	164.7	1.12	57.3	1,972
2007	—	—	51.0	172.6	1.18	60.0	2,103
2008	—	—	51.0	—	1.18	60.0	2,239

Source: FAO State of Marine Fisheries (year for which stock status is available indicated in bold).

Note: Section 4.1. in the main text indicates how the calculations in this table were made. — Data are not available. Base years for indexes are shown in italic.

a. Because the FAO's assessment of the state of marine fish stocks is not available for certain years, values from preceding year are used.

b. The deflator is that used by the U.S. Labor Department for all commodities.

Figure A4.1   Example of Increasing Rents in New Zealand
and Icelandic Fisheries

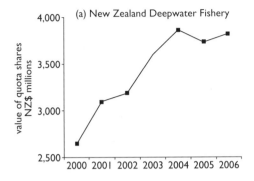

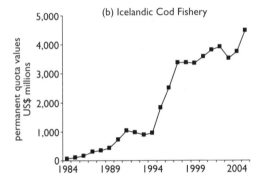

Source: (a) PROFISH Team, World Bank, based on New Zealand deepwater fishery monetary
stock accounts; (b) Authors.
Note: The quota share value is used as a proxy for rents.

REFERENCES

Agnarsson, S., and R. Arnason. 2007. "The Role of the Fishing Industry in the Icelandic Economy." In *Advances in Fisheries Economics,* ed. T. Bjorndal, D. V. Gordon, R. Arnason, and U. R. Sumaila. Oxford: Blackwell Publishing.

Ahrens, R., and C. Walters. 2005. "Why Are There Still Large Pelagic Predators in the Oceans? Evidence of Severe Hyper-depletion in Longline Catch-per-Effort." Paper presented at first meeting of the Scientific Committee of the Western and Central Pacific Fisheries Commission, Noumea, New Caldonia.

Alchian, Armen A. 1987. "Rent." In *The New Palgrave Dictionary of Economics,* eds. Steven N. Durlauf and Lawrence E. Blume. 2d ed. New York: Palgrave Macmillan, 2008. http://www.dictionaryofeconomics.com/article?id=pde2008_R000098.

Anderson, C. N. K., C.-H Hsieh, S. A. Sandin, R. Hewitt, A. B. Hollowed, J. Beddington, R. M. May, and G. Sugihara. 2008. "Why Fishing Magnifies Fluctuations in Fish Abundance." *Nature* 452 (7189): 835–39.

APFIC (Asia-Pacific Fisheries Commission). 2006. "Selected Issues of Regional Importance: Low Value/Trash Fish." Twenty-ninth APFIC Session. Kuala Lumpur, August 21–23.

Arnason, R. 1984. "Efficient Harvesting of Fish Stocks: The Case of the Icelandic Demersal Fisheries." PhD diss. University of British Columbia.

———. 1995. *The Icelandic Fisheries: Evolution and Management of a Fishing Industry.* Oxford: Fishing News Books.

———. 2006. "Estimation of Global Rent Loss in Fisheries: Theoretical Basis and Practical Considerations." In *Proceedings of the Thirteenth Biennial Conference of the International Institute of Fisheries Economics and Trade,* ed. A. Shriver. Corvallis, OR: IIFET.

———. 2007. "An Excel Program to Estimate Parameters and Calculate Fisheries Rents." Food and Agriculture Organization, Rome.

————. 2008. "Iceland's ITQ System Creates New Wealth." *Electronic Journal of Sustainable Development* 1 (2). www.ejsd.org.

Arnason, R., R. Hannesson, and W. E. Schrank. 2000. "Costs of Fisheries Management: The Cases of Iceland, Norway and Newfoundland." *Marine Policy* 24 (3): 233–43.

Asche, F., and T. Bjørndal. 1999. *Demand Elasticities for Fish: A Review.* Globefish Special Series 9. Food and Agriculture Organization, Rome.

Asche F., T. Bjørndal, and D. V. Gordon. 2003. "Fishermen Behaviour with Individual Vessel Quotas—Overcapacity and Potential Rent." Paper presented at the Fifteenth Annual Conference of the European Association of Fisheries Economists. French Research Institute for Exploitation of the Sea, Issy-les-Moulineaux.

Baird, Andrew, J. A. Maynard, O. Hoegh-Guldberg, P. J. Mumby, A. J. Hooten, R. S. Steneck, P. Greenfield, E. Gomez, D. R. Harvell, P. F. Sale, A. J. Edwards, K. Caldeira, N. Knowlton, C. M. Eakin, R. Iglesias-Prieto, N. Muthiga, R. H. Bradbury, A. Dubi, and M. E. Hatziolos, 2008. "Coral Adaptation in the Face of Climate Change." *Science* 320 (5874): 315–16.

Bertignac, Michel, H. F. Campbell, J. Hampton, and A. J. Hand. 2001. "Maximizing Resource Rent from the Western and Central Pacific Tuna Fisheries." *Marine Resource Economics* 15: 151–77.

Björndal, T. 1987. "Production Economics and Optimal Stock Size in a North Atlantic Fishery." *Scandinavian Journal of Economics* 89: 145–64.

Blaug, M. 2000. "Henry George: Rebel with a Cause." *European Journal of the History of Economic Thought* 7: 270–88.

Cesar, H. 1996. "Economic Analysis of Indonesian Coral Reefs." Environment Department, World Bank, Washington, DC.

————. 2000. "Collected Essays on the Economics of Coral Reefs." Cordio. Department for Biology and Environmental Sciences, Kalmar University, Sweden.

Christy, F. T. 1999. "Common Property Rights: An Alternative to ITQs." Paper prepared for the FAO/Western Australia Fishrights Conference on the Use of Property Rights in Fisheries Management, November 11–19. Fremantle, Australia.

Christy, F. T., and A. Scott. 1965. *The Common Wealth in Ocean Fisheries. Some Problems of Growth and Economic Allocation.* Baltimore, MD: Johns Hopkins University Press for Resources for the Future.

Chuenpagdee, R., Lisa Liguori, Maria L. D. Palomares, and Daniel Pauly. 2006. "Bottom-Up, Global Estimates of Small-Scale Marine Fisheries Catches." Fisheries Centre Research Reports 14 (8). Fisheries Centre, University of British Columbia, Vancouver.

Clark, C. W. 1976. *Mathematical Bioeconomics: The Optimal Management of Renewable Resources.* Hoboken, NJ: John Wiley & Sons.

Clark, C. W. and G. R. Munro. 1975. "The Economics of Fishing and Modern Capital Theory: A Simplified Approach." *Journal of Environmental Economics and Management* 2: 92–106.

Clark, C. W., G. R. Munro, and U. R. Sumaila. 2007. "Buyback Subsidies, the Time Consistency Problem, and the ITQ Alternative." *Land Economics* 83 (1): 50–58.

Coglan, L., and S. Pascoe. 1999. "Separating Resource Rents from Intra-marginal Rents in Fisheries: Economic Survey Data." *Agricultural and Resource Economics Review* 28: 219–28.

Commission of the European Communities, International Monetary Fund, Organisation for Economic Co-operation and Development, United Nations, and World Bank. 1993. System of National Accounts 1993. Prepared under the auspices of the

Inter-Secretariat Working Group on National Accounts. http://unstats.un.org/unsd/sna1993/introduction.asp.

Committee to Review Individual Fishing Quotas, Ocean Studies Board, Commission on Geosciences, Environment, and Resources, and National Research Council. 1999. *Sharing the Fish: Toward a National Policy on Individual Fishing Quotas.* Washington, DC: National Academy Press.

Concerted Action: Economic Assessment of European Fisheries. 2004. "Economic Performance of Selected European Fishing Fleets." Q5CA-2001-01502. Commission of the European Communities.

Corveler, Tangi. 2002. *Illegal, Unreported and Unregulated Fishing, Taking Action for Sustainable Fisheries.* Wellington: WWF New Zealand.

Costello, C., S. D. Gaines, and J. Lynham. 2008. "Can Catch Shares Prevent Fisheries Collapse?" *Science* 321 (September 19).

Cunningham, S., and T. Bostock, eds. 2005. *Successful Fisheries Management: Issues, Case Studies, Perspectives.* Delft: Eburon Publishers.

Curtis, R., and D. Squires, eds. 2007. *Fisheries Buybacks.* Oxford: Blackwell.

Dalton, R. 2005. "Conservation Policy: Fishy Futures." *Nature* 437 (September 22): 473–74.

Danielsson, A. 2005. "Methods for Environmental and Economic Accounting for the Exploitation of Wild Fish Stocks and Their Application to the Case of Icelandic Fisheries." *Environmental and Resource Economics* 31: 405–30.

Dasgupta, P. S., and G. M. Heal. 1979. *Economic Theory and Exhaustible Resources.* Cambridge, U.K.: James Nisbet and Cambridge University Press.

Davidson, R., and J. G. MacKinnon. 1993. *Estimation and Inference in Econometrics.* Oxford: Oxford University Press.

Davidsson, K. 2007. Cited by G. Valdimarsson in "Fish in the Global Food Supply Chain." Paper presented at the World Seafood Congress, Dublin, September 25–27.

de Soto, H. 2000. *The Mystery of Capital: Why Capitalism Triumphs in the West and Fails Everywhere Else.* New York: Random House.

Delgado, C. L., N. Wasa, M. W. Rosegrant, S. Meijer, and A. Mahfuzuddin. 2003. *Fish to 2020. Supply and Demand in Changing Global Markets.* Washington, DC: International Food Policy Research Institute/World Fish Centre.

Department of the Environment, Water, Heritage and the Arts, Australia. 2008. Fisheries and Environment. http://www.environment.gov.au/coasts/fisheries/index.html.

DFID (UK Department for International Development). 2004. "Resource Rents: Fiscal Reform in Fisheries." www.keysheets.org.

Dupont, D. P. 1990. "Rent Dissipation in Restricted Access Fisheries." *Journal of Environmental Economics and Management* 19 (1): 26–44.

Edwards, S. F., and S. J. Murawski. 1993. "Potential Economic Benefits from Efficient Harvest of New England Groundfish." *North American Journal of Fisheries* 13 (3): 437–49.

Energy Information Agency, U. S. Department of Energy. 2007. "World Crude Oil Price." http://tonto.eia.doe.gov/dnav/pet/pet_pri_wco_k_w.htm.

Essington, T. E., A. H. Beaudreau, and J. Weidenmann. 2006. "Fishing through Marine Food Webs." *Proceedings of the National Academy of Sciences* (USA) 103: 3171–75.

European Commission. 2007. "Climate Change: What Impact on Fisheries?" *Fisheries and Aquaculture in Europe.* 35 (August): 4–8.

FAO (Food and Agriculture Organization). 1993. "Marine Fisheries and the Law of the Sea: A Decade of Change." Special chapter (revised) of *The State of Food and Agriculture 1992*. FAO Circular 853. Rome.

——. 1995. "Code of Conduct for Responsible Fisheries." Rome.

——. 1999a. "Numbers of Fishers." FAO Fisheries Circular 929, Rev. 2. Rome.

——. 1999b. *The State of World Fisheries and Aquaculture 1998*. Rome: FAO.

——. 2000. "Report of the Technical Consultation on the Measurement of Fishing Capacity. Mexico City, 29 November –3 December." FAO Fisheries Report 615. Rome.

——. 2001a. "International Plan of Action to Prevent, Deter and Eliminate IUU Fishing (IPOA-IUU)." Rome.

——. 2001b. "Techno-economic performance of marine capture fisheries." FAO Fisheries Technical Papers 421. Rome.

——. 2002. *The State of World Fisheries and Aquaculture 2002*. Rome.

——. 2004a. *Integrated Environmental and Economic Accounting for Fisheries*. Rome: FAO.

——. 2004b. *The State of World Fisheries and Aquaculture 2004*. Rome: FAO.

——. 2005. "Review of the State of World Marine Fishery Resources." FAO Fisheries Technical Papers 457. Rome.

——. 2006. "The State of World Highly Migratory, Straddling and Other High Seas Fishery Resources and Associated Species." FAO Fisheries Technical Papers 495. Rome.

——. 2007a. "Fish and Fishery Products: World Apparent Consumption Statistics Based on Food Balance Sheets (1961–2003)." Fisheries Circular 821, Rev. 8. Rome.

——. 2007b. "Increasing the Contribution of Small-Scale Fisheries to Poverty Alleviation and Food Security." FAO Fisheries Technical Paper 481. Rome.

——. 2007c. *The State of World Fisheries and Aquaculture 2006*. Rome: FAO.

——. 2008. "Climate Change for Fisheries and Aquaculture." Technical Background Document from the Expert Consultation held on April, 7–9, Rome.

——. Various years. FAO Food Balance Sheets. http://faostat.fao.org/site/368/default.aspx.

——. Various years. Fish Stat Plus. http://www.fao.org/fishery/statistics/software/fishstat.

Fishman, G. S. 1996. *Monte Carlo: Concepts, Algorithms, and Applications*. New York: Springer.

Fisman, R., and E. Miguel. 2006. "Cultures of Corruption: Evidence from Diplomatic Parking Tickets." NBER Working Paper 12312. National Bureau of Economic Research, Cambridge, MA. http://www.nber.org/papers/w12312 .

Fitzpatrick, J., 1996. " Technology and Fisheries Legislation." In *Precautionary Approach to Fisheries*. Part 2: *Scientific Papers*. FAO Fisheries Technical Paper 350, Part 2. Rome: FAO.

Fox, W. W. 1970. "An Exponential Surplus Model for Optimizing Exploited Fish Populations." *Transactions of the American Fisheries Society* 99: 80–88.

Garcia, Serge M., and Richard J. R. Grainger. 2005. "Gloom and Doom? The Future of Marine Capture Fisheries." *Phil. Trans. R. Soc. B* (2005) 360: 21–46. Published online, January 29, 2005.

Garcia, S. M., and C. Newton. 1997. "Current Situation, Trends and Prospects in World Capture Fisheries." In *Global Trends: Fisheries Management,* ed. E. L. Pickitch, D. D. Huppert, and M. P. Sissenwine. Symposium 20. Bethesda, MD: American Fisheries Society.

Gillett, R., and C. Lightfoot. 2002. *The Contribution of Fisheries to the Economies of Pacific Island Countries.* Manila: Asian Development Bank, Pacific Studies Series.

Gordon, H. S. 1954. "The Economic Theory of a Common Property Resource: The Fishery." *Journal of Political Economy* 62: 124–42.

Grafton, R. Q., R. Hilborn, L. Ridgeway, D. Squires, M. Williams, S. Garcia, T. Groves, J. Joseph, K. Kelleher, T. Kompas, G. Libecap, C. G. Lundin, M. Makino, T. Matthiasson, R. McLoughlin, A. Parma, G. San Martin, B. Satia, C.-C. Schmidt, M. Tait, and L. X. Zhang, 2008. "Positioning Fisheries in a Changing World. *Marine Policy* 32 (4): 630–34.

Grafton, R. Q., T. Kompass, and R. W. Hilborn. 2007. "Economics of Overexploitation Revisited." *Science* 318 (December 7).

Gulland, J. A. 1971. *The Fish Resources of the Ocean.* West Byfleet, UK: Fishing News (Books) Ltd.

Hannesson, R. 1986. "The Economic Characteristics of the Management of the Inshore Fishery in Cyrpus." FAO, Rome.

———. 1993. "Bioeconomic Analysis of Fisheries." FAO, Rome.

———. 2002. "The Economics of Fishing Down the Food Chain." *Canadian Journal of Fisheries and Aquatic Sciences* 59 (5): 755–58.

Hardin G. 1968. "The Tragedy of the Commons." *Science* 162:1243–48.

Herrmann, M. 1996. "Estimating the Induced Price Increase for Canadian Pacific Halibut with the Introduction of the Individual Vessel Quota System." *Canadian Journal of Agricultural Economics* 44: 151–64.

High Seas Task Force. 2006. *Closing the Net: Stopping Illegal Fishing on the High Seas.* Governments of Australia, Canada, Chile, Namibia, New Zealand, and the United Kingdom, World Wildlife Fund, IUCN (International Union for Conservation of Nature), and the Earth Institute at Columbia University.

Homans, F. R., and J. E. Wilen. 1997. "A Model of Regulated Open Access Resource Use." *Journal of Environmental Economics and Management* 32: 1–21.

———. 2005. "Markets and Rent Dissipation in Regulated Open Access Fisheries." *Journal of Environmental Economics and Management* 49: 381–404.

Hoshino, E., and Y. Matsuda. 2007. "Resource Rents in two Japanese Fisheries." Final draft prepared for the World Bank PROFISH Program, Washington, DC.

ICTSD (International Centre for Trade and Sustainable Development). 2006. "Fisheries, International Trade and Sustainable Development: Policy Discussion Paper." ICTSD Natural Resources, International Trade and Sustainable Development Series, Geneva.

ILO (International Labour Organization). 2000. "Note on the Proceedings of a Tripartite Meeting on Safety and Health in the Fishing Industry, December 13–17," Geneva.

IPCC (Intergovernmental Panel on Climate Change). 2007. *Climate Change 2007: Synthesis Report. Contribution of Working Groups I, II and III to the Fourth Assessment. Report of the Intergovernmental Panel on Climate Change,* ed. R. K. Pachauri and A. Reisinger. Geneva: IPCC.

Jensen R. 2007. "The Digital Provide: Information (Technology), Market Performance, and Welfare in the South Indian Fisheries Sector." *Quarterly Journal of Economics* 122 (3): 879–924.

Johannes, R. E. 1978. "Traditional Marine Conservation Methods in Oceania and Their Demise." *Annual Review of Ecology and Systematics* 9: 349–64.

Josupeit, H. 2008. FAO presentation on commodity trade development to a GLOBEFISH partner meeting, Brussels, April 21.

Kelleher, K. 2002a. "The Costs of Monitoring, Control and Surveillance of Fisheries in Developing Countries." FAO Fisheries Circular 976. FAO, Rome.

———. 2002b. "Robbers, Reefers and Ramasseurs. A Review of Selected Aspects of Fisheries Monitoring Control and Surveillance in Seven West African Countries." Sub-Regional Fisheries Commission. Project FAO/GCP/INT/722/LUX (AFR/013). July.

———. 2005. "Discards in the World's Marine Fisheries. An Update." FAO Fisheries Technical Paper 470. FAO, Rome.

Kelleher, K., and R. Willmann. 2006. "'The Rent Drain': Towards an Estimate of the Loss of Resource Rents in the World's Fisheries." Report of the FAO–World Bank Study Design Workshop, January 17–18, World Bank, Washington, DC.

Kirkley, J. E., J. M. Ward, J. Nance, F. Patella, K. Brewster-Geisz, C. Rogers, E. Thunberg, J. Walden, W. Daspit, B. Stenberg, S. Freese, J. Hastie, S. Holiman, and, M. Travis. 2006. "Reducing Capacity in U.S. Managed Fisheries." NOAA Technical Memorandum NMFS-F/SPO-76. National Oceanic and Atmospheric Administration, Washington, DC.

Kjorup, M. 2007. Presentation of the National Fisheries Policies Office of the Danish Food, Agriculture and Fisheries Ministry to the OECD Committee on Fisheries, October, Paris.

Kurien, J. 2007. "Estimation of Some Aspects of the Economics of Operation of Marine Fishing Crafts in India." A study for the PROFISH Big Numbers Project. WorldFish and FAO, Penang and Rome.

Lery, J.-M., J. Prado, and U. Tietze. 1999. "Economic Viability of Marine Capture Fisheries: Findings of a Global Study and an Interregional Workshop." FAO Fisheries Technical Paper 377. FAO, Rome.

Lloyd's Register – Fairplay. http://www.lrfairplay.com/.

Lutz, M. 2008. "Overfishing Has Reduced Ocean CO_2 Sequestration." Powerpoint presentation to PROFISH Steering Committee. World Bank, Washington, DC.

Milazzo, M. 1998. "Subsidies in World Fisheries: A Reexamination." World Bank Technical Paper 406. Washington, DC.

Millennium Ecosystem Assessment. 2005. *Ecosystems and Human Well-Being: Synthesis.* Washington, DC: Island Press. http://www.millenniumassessment.org/en/index.aspx.

MRAG and UBC Fisheries Centre. 2008. *The Global Extent of Illegal Fishing.* Final Report. Marine Resources Assessment Group and University of British Columbia Fisheries Centre, April 28, Vancouver.

Myers, R. A., and B. Worm. 2003. "Rapid Worldwide Depletion of Predatory Fish Communities." *Nature* 423: 280–83.

Nguyen, L., and T. B. Nguyen. 2008. "Assessment of Tonkin Gulf Fishery, Vietnam, Based on Bioeconomic Models." Paper presented to the Fourtheenth Biennial Conference of the International Institute of Fisheries Economics and Trade, Nha Trang, Vietnam.

Nielsen, M., B. Cozzari, G. Eriksen, O. Flaaten, E. Gudmundsson, J. Løkkegaard, K. Petersen, and S. Waldo. 2007. "Focus on the Economy of the Nordic Fisheries. Case Study Reports from Iceland, Norway, the Faroe Islands, Sweden and Denmark." Report 186. Institute of Food and Resource Economics, Bonn.

NMFS (National Marine Fisheries Service). 2008. Fish Stock Sustainability Index (FSSI). NOAA, Washington, DC. http://www.nmfs.noaa.gov/sfa/domes_fish/StatusoFisheries/2008/1stQuarter/Q1-2008-FSSISummaryChanges.pdf.

Oceanic Développement. 2001. "Cost Benefit Comparison of Different Control Strategies." Final report, January, prepared in association with Richard Banks Ltd., Megapesca Lda.

OECD (Organisation for Economic Co-operation and Development). 2000. "Transition to Responsible Fisheries. Government Financial Transfers and Resource Sustainability: Case Studies." AGR/FI(2000)10/FINAL. Paris.

———. 2006. "Why Fish Piracy Persists. The Economics of Illegal, Unreported and Unregulated Fishing." Paris.

———. 2008. "Reforming Fisheries Policies: Insights from the OECD Experience." Trade and Agriculture Directorate Fisheries Committee TAD/FI(2008)4. Paris.

Pan Chenjun. 2007. "Overview of China's Seafood Industry." Rabobank Seafood Industry Roundtable, October, Bangkok.

Paredes, C. 2008. "La industria anchovetera Peruana: Costos y Beneficios. Un Análisis de su Evolución Reciente y de los Retos par el Futuro." Estudio preparado por encargo del Banco Mundial al Instituto del Perú de la Universidad de San Martin de Porres. Trabalho en processo.

Pauly, D. 1995. "Anecdotes and the Shifting Baseline Syndrome of Fisheries." *Trends in Ecology and Evolution* 10 (10): 430.

Pauly, D., V. Christensen, J. Dalsgaard, R. Froese, and F. Torres, Jr. 1998. "Fishing Down Marine Food Webs." *Science* 279: 860–63.

Pauly, D., V. Christensen, S. Guénette, Tony J. Pitcher, U. R. Sumaila, C. J. Walters, R. Watson, and D. Zeller. 2002. "Towards Sustainability in World Fisheries." *Nature* 418 (August 8): 689–95.

Pitcher, T. R., R. Watson, R. Forrest, H. Valtýsson, and Sylvie Guénette. 2002. "Estimating Illegal and Unreported Catches from Marine Ecosystems: A Basis for Change." *Fish and Fisheries.* 3 (4): 317–39.

Pricewaterhouse Coopers LLP. 2000. "Study into the Nature and Extent of Subsidies in the Fisheries Sector of APEC Members Economies." APEC 00-FS-01.1. Asia Pacific Economic Cooperation, Singapore.

Repetto, R. 2002. "Creating Asset Accounts for a Commercial Fishery Out of Equilibrium: A Case Study of the Atlantic Sea Scallop Fishery." *Review of Income and Wealth* 48: 245–59.

Ricardo, D. 1817 (1951). "Principles of Political Economy and Taxation." In *The Works and Correspondence of David Ricardo,* ed. P. Sraffa and M. Dobb. Cambridge, UK: Cambridge University Press.

Richard, G., and D. Tait. 2007. "Shrimp Twin Trawl Technology." Minister of Public Works and Government Services, Canada. DFO/5516. Cat. No. Fs 23-300/2-1997E.

Salz, P. 2006. "Economic Performance of EU Fishing Fleets and Consequences of Fuel Price Increase." Contribution to the Conference on Energy Efficiency in Fisheries, Brussels, May 11–12.

Sanchirico, J. N., and J. E. Wilen. 2002. "Global Marine Fisheries Resources: Status and Prospects." Sustainable Development Issue Brief 02–17. Resources for the Future, Washington, DC.

Sanchirico, J. N., and J. E. Wilen. 2007. "Global Marine Fisheries Resources: Status and Prospects." *International Journal of Global Environmental Issues* 7 (2–3): 106–18.

Schaefer, M. B. 1954. "Some Aspects of the Dynamics of Populations Important to the Management of Commercial Marine Species." *Inter-American Tropical Tuna Commission Bulletin* 1: 27–56.

Schorr, D. K., and J. F. Caddy. 2007. "Sustainability Criteria for Fisheries Subsidies Options for the WTO and Beyond." Prepared for the United Nations Environment Programme, Economics and Trade Branch, and World Wide Fund for Nature (WWF). Geneva.

Schrank, W. E. 2003. "Introducing Fisheries Subsidies." FAO Fisheries Technical Paper 437. FAO, Rome.

Scott, A. D. 1955. "The Fishery: The Objectives of Sole Ownership." *Journal of Political Economy* 63: 116–24.

Seafood International. 2008. Supplies and Markets. May 2008.

Sharp, P. and U. R. Sumaila. Forthcoming. "Quantification of U.S. Marine Fisheries Subsidies." *North American Journal of Fisheries Management.*

Shotton, R., ed. 1999. *Use of Property Rights in Fisheries Management.* vol. 2. FAO Fisheries Technical Paper 404/2. Rome: FAO.

Sibert, J., J. Hampton, P. Kleiber, and M. Maunder. 2006. "Biomass, Size, and Trophic Status of Top Predators in the Pacific Ocean." *Science* 314 (5806): 1773–76. www.sciencemag.org/cgi/content/full/314/5806/1773/DC1

Smith, A. 1776 (1981). *An Inquiry into the Nature and Causes of the Wealth of Nations,* ed. R. H. Cambell and A. J. Skinner. Indianapolis: Liberty Fund.

Smith, A. 2009. In press. Untitled manuscript of report on energy for fisheries prepared for FAO.

Southwick Associates. 2005. "The Economics of Recreational and Commercial Striped Bass Fishing in Maryland." Report prepared for Stripers Forever, Inc., by Southwick Associates, Fernandina Beach, FL, December.

———. 2006. "The Relative Economic Contributions of U.S. Recreational and Commercial Fisheries." Report prepared for the Theodore Roosevelt Conservation Partnership by Southwick Associates, Fernandina Beach, FL, April.

Sumaila, U. R., J. Alder, and H. Keith. 2006. "Global Scope and Economics of Illegal Fishing." *Marine Policy* 30 (6): 696–703.

Sumaila, U. R., and D. A. Marsden. 2007. "Case Study of the Namibian Hake Fishery." Prepared for the FAO/World Bank rent drain study. Fisheries Economics Research Unit, Fisheries Centre, University of British Columbia, Vancouver.

Sumaila, U. R., and L. Suatoni. 2006. "Economic Benefits of Rebuilding U.S. Ocean Fish Populations." Fisheries Centre, University of British Columbia, Vancouver.

Sumaila, U. R., L. Teh, R. Watson, P. Tyedmers, and D. Pauly. 2008. "Fuel Price Increase, Subsidies, Overcapacity, and Resource Sustainability." *ICES Journal of Marine Science* 65 (September): 832–40.

Sumaila, U.R., and D. Pauly, eds. 2006. "Catching More Bait: A Bottom-Up Re-Estimation of Global Fisheries Subsidies." Fisheries Centre Research Reports 2006, 14 (6). Fisheries Centre, University of British Columbia, Vancouver.

Sustainable Fisheries Livelihoods Programme. 2007. "Building Adaptive Capacity to Climate Change: Policies to Sustain Livelihoods and Fisheries." Policy Brief on Development Issues 8. FAO, Rome. http://www.sflp.org/briefs/eng/policybriefs. html.

Sutinen, J. G., and K. Kuperan. 1994. *A Socioeconomic Theory of Regulatory Compliance in Fisheries*. Proceedings of the VIIth Conference of the International Institute of Fisheries Economics and Trade, Taipei, July 18–21.

Tacon, A. 2006. "Thrash Fish Fisheries, Aquaculture, Pellets and Fishmeal Substitutes." Prepared for the Asia-Pacific Fisheries Commission, Regional Consultative Forum, Kuala Lumpur.

Tai, S. Y., and T. Heaps. 1996. "Effort Dynamics and Alternative Management Policies for the Small Pelagic Fisheries of Northwest Peninsular Malaysia." *Marine Resource Economics* (11): 1–19.

Tietze, U., J. Prado, J-M. Le Ry, and R. Lasch. 2001. "Techno-Economic Performance of Marine Capture Fisheries and the Role of Economic Incentives, Value Addition and Changes of Fleet Structure. Findings of a Global Study and an Interregional Workshop." FAO Fisheries Technical Paper 421. FAO, Rome.

Tietze, U., W. Thiele, R. Lasch, B. Thomsen, and D. Rihan. 2005. "Economic Performance and Fishing Efficiency of Marine Capture Fisheries." FAO Fisheries Technical Paper 482. FAO, Rome.

Townsend, R. 1990. "Entry Restrictions in the Fishery: A Survey of the Evidence." *Land Economics* 66 (4).

Tyedmers, P. H., R. Watson, and D. Pauly. 2005. "Fueling Global Fishing Fleets." *Ambio* 34 (8): 636–38.

UN and FAO. 2004. "Handbook of National Accounting: Integrated Environmental and Economic Accounting for Fisheries." United Nations and Food and Agriculture Organization.

UNEP (United Nations Environmental Programme) World Conservation Monitoring Centre. 2006. "In the Front Line: Shoreline Protection and Other Ecosystem Services from Mangroves and Coral Reefs." Cambridge, UK.

U.S. Bureau of Labor Statistics. 2007. http://www.bls.gov.

Varian, H. 1984. *Microeconomic Theory*. 2d. ed. New York: W.W. Norton & Company.

Volterra, V. 1926. "Fluctuations in the Abundance of a Species Considered Mathematically." *Nature* 118: 558–60.

Ward, J. M. 2006. "Dissipation of Resource Rent in the Gulf of Mexico Shrimp Fishery." U.S. National Marine Fisheries Service, Washington, DC.

Warui, S. W. 2008. "Rents and Rents Drain in the Lake Victoria Nile Perch Fishery." Ministry of Livestock and Fisheries Development, Kenya, and University of Iceland/United Nations University.

Watson, J. M., and R. Seidel. 2003. "2003 Economic Survey of the North Sea and West of Scotland Whitefish Fleet." Report for Seafish, Grimsby, Scotland (December).

Watson, R., and D. Pauly. 2001. "Systematic Distortions in World Fisheries Catch Trends." *Nature* 414 (November 29): 534–36.

WHAT (Humanity Action Trust). 2000. "Governance for a Sustainable Future." London. www.what.org.uk.

Wilen, J. E. 2005. "Property Rights and the Texture of Rents in Fisheries." In *Evolving Property Rights in Marine Fisheries*, ed. D. R. Leal. Lanham, MD: Rowman and Littlefield.

Wilen, J. E., and E. J. Richardson. 2003. "The Pollock Conservation Cooperative." Paper prepared for Workshop on Cooperatives in Fisheries Management, June, Anchorage Alaska.

Willmann, R. 2000. "Group and Community-Based Fishing Rights." In *Use of Property Rights in Fisheries Management: Proceedings of the FishRights99 Conference,* ed. R. Shotton, pp. 51–57. FAO Fisheries Technical Paper 404/1. Rome: FAO.

Willmann, R., P. Boonchuwong, and S. Piumsombun. 2003. "Fisheries Management Costs in Thai Marine Fisheries." In *The Cost of Fisheries Management,* ed. W. E. Schrank, R. Arnason, and R. Hannesson, pp. 187–219. Farnham, UK: Ashgate Publishing Limited.

World Bank. 2005. *Where Is the Wealth of Nations. Measuring Capital for the 21st Century.* Washington, DC: World Bank. http://siteresources.worldbank.org/INTEEI/214578-1110886258964/20748034/All.pdf.

———. 2007a. *Changing the Face of the Waters. The Promise and Challenge of Sustainable Aquaculture.* Washington DC: World Bank.

World Bank, 2007b. "Strengthening World Bank Group Engagement on Governance and Anticorruption." World Bank, Washington, DC.

World Bank, 2007c. *World Development Report 2008: Agriculture for Development.* Washington, DC 2007.

World Bank. 2008. "Rising Food Prices: Policy Options and World Bank Response." http://siteresources.worldbank.org/NEWS/Resources/risingfoodprices_backgroundnote_apr08.pdf.

World Bank and IUCN (International Union for Conservation of Nature). Forthcoming. Report of the PROFISH/ IUCN Workshop on Corruption in Fisheries. World Bank, Washington, DC.

World Summit on Sustainable Development. 2002. Plan of Implementation Johannesburg 2002. http://www.un.org/esa/sustdev/documents/WSSD_POI_PD/English/WSSD_PlanImpl.pdf.

Worm B., E. B. Barbier, N. Beaumont, J. E. Duffy, C. Folke, B. S. Halpern, J. B. C. Jackson, H. K. Lotze, F. Micheli, S. R. Palumbi, E. Sala, K. Selkoe, J. J. Stachowicz, and R. Watson. 2006. "Impacts of Biodiversity Loss on Ocean Ecosystem Services." *Science* 314: 787–90.

WTO (World Trade Organization). 1994. Agreement on Subsidies and Countervailing Measures. World Trade Organization. Geneva. http://www.wto.org/english/docs_e/legal_e/24-scm.pdf.

Zeller D., and D. Pauly, eds. 2007. "Reconstruction of Marine Fisheries Catches for Key Countries and Regions (1950–2005)." Fisheries Centre Research Reports 15 (2). Fisheries Centre, University of British Columbia, Vancouver.

Zeller, D., S. Booth, and D. Pauly. 2006. "Fisheries Contributions to the Gross Domestic Product: Underestimating Small-Scale Fisheries in the Pacific." *Marine Resource Economics* 21: 355–74.

INDEX